Lecture Notes in Statistics 84

Edited by S. Fienberg, J. Gani, K. Krickeberg, I. Olkin, and N. Wermuth

Sakutarō Yamada

Pivotal Measures in Statistical Experiments and Sufficiency

Springer-Verlag
New York Berlin Heidelberg London Paris
Tokyo Hong Kong Barcelona Budapest

Sakutarō Yamada
Department of Fishery Resources Management
Tokyo University of Fisheries
Konan 4-5-7, Minato-ku, Tokyo 108
Japan

Yamada, Sakutarō.
Pivotal measures in statistical experiments and sufficiency/
Sakutarō Yamada.
p. cm. -- (Lecture notes in statistics)
Includes bibliographical references and index.
ISBN-13:978-0-387-94216-2
1. Experimental design. 2. Estimation theory. I. Title.
II. Series: Lecture notes in statistics (Springer-Verlag)
QA279.Y36 1994
519.5--dc20 93-47440

Printed on acid-free paper.

Camera ready copy provided by the author.

9 8 7 6 5 4 3 2 1

ISBN-13:978-0-387-94216-2 e-ISBN-13:978-1-4612-2644-4
DOI: 10.1007/978-1-4612-2644-4

Preface

In the present work I want to show a mathematical study of the statistical notion of sufficiency mainly for undominated statistical experiments. The famous Burkholder's (1961) and Pitcher's(1957) examples motivated some researchers to develop new theory of sufficiency. Le Cam (1964) is probably the most excellent paper in this field of study.

This note also belongs to the same area. Though it is more restrictive than Le Cam 's paper(1964), a study which is connected more directly with the classical papers of Halmos and Savage(1949), and Bahadur(1954) is shown. Namely I want to develop a study based on the notion of pivotal measure which was introduced by Halmos and Savage(1949).

It is great pleasure to have this opportunity to thank Professor H. Heyer and Professor H. Morimoto for their careful reading the manuscript and valuable comments on it. I am also thankful to Professor H. Luschgy and Professor D. Mussmann for their proposal of writing "the note".

I would like to dedicate this note to the memory of my father Eizō.

Tokyo, August 1993
Sakutarō Yamada

Contents

Chapter 0. Introduction

In these notes we present a theory of sufficiency which covers undominated statistical experiments as well as dominated ones. The familiar topics in the dominated case, such as pairwise sufficiency, Neyman factorization, minimal sufficient statistics, the Rao-Blackwell theorem, are treated from a more general view point than in the Halmos-Savage-Bahadur scheme and sometimes in a slightly different way, while the well known "pathologies" of Pitcher (1957) and Burkholder (1961) are averted.

The concepts of pivotal measure and PSS (pairwise sufficiency with supports) play a fundamental role in this theory. The former is an extension of the "pivotal measure" in the dominated case, a special type of dominating measure defined by Halmos and Savage (1949) and called as such in Bahadur (1957). It acted as the key concept in their proofs of the Neyman factorization theorem and in the construction of the minimal sufficient subfield. The Neyman factorization theorem in turn implied that in the dominated case a subfield including a sufficient subfield is itself sufficient. Our extended concept will have applications of the same nature to the undominated cases.

PSS, on the other hand, is a concept which is weaker than sufficiency and stronger than pairwise sufficiency. Needless to say that in the dominated case all these three concepts coincide.

The well known difficulty concerning sufficiency in the undominated cases is that a minimal sufficient subfield may not exist (Pitcher pathology), and a non-sufficient subfield can include a sufficient one (Burkholder pathology). It will be shown in later chapters that if a pivotal measure is defined in connection with PSS, not with sufficiency, then a Neyman factorization emerges as a criterion of PSS, a subfield including a PSS subfield is itself PSS, and there exists a minimal PSS subfield. Thus PSS appears as a generalization and a kind of replacement of sufficiency in our treatment of the undominated cases.

The foregoing description would show that we pursue the measure-theoretic approach of earlier studies and try to develop further. This is in contrast with the work of Le Cam (1964), probably the most outstanding attempt at constructing a theory of sufficiency free from above-mentioned pathologies, which adopts the theory of vector lattice as its framework.

For a recent survey on sufficiency we refer to Yamada and Morimoto(1992).

In Chapter 1 we give a few types of undominated statistical experiments and discuss some of their basic properties. Our attension is mainly focused on two types of experiments. They are the majorized experiments and the weakly dominated experiments. The former type concerns the most general experiments treated in this note. Though some non-majorized experiments do appear, they are mainly used to construct counter examples.

Majorized experiments are treated in Section 1.1. They are defined as statistical experiments $\mathcal{E}=(\mathcal{X},\mathcal{A},\mathcal{P}=\{P_\theta;\theta\in\Theta\})$ with a family $\mathcal{P}$ of probability measures on $(\mathcal{X},\mathcal{A})$ such that there exists a "majorizing measure" m, w.r.t. which each P_θ in $\mathcal{P}$ has a density. Each P_θ has thus a support, the part of $\mathcal{X}$ on which its density is positive. It will be shown that a support can be defined in a more general context without direct reference to the majorizing measures or densities, and if each P_θ has a support in this sense then $\mathcal{E}$ is majorized.

The concept of support is then applied to construct a maximal decomposition of the experiment, which in turn is used to construct a majorizing measure equivalent to $\mathcal{P}$. An equivalent majorizing measure, as we call it, plays an important role in the later chapters.

Section 1.2 treats weakly dominated experiments. They are defined as majorized experiments whose majorizing measure can be taken as a localizable one. We start by giving some measure theoretic results on localizable measures and prove that a weakly dominated experiment has an equivalent localizable majorizing measure. We then show that what is called coherent experiments are the same as the weakly dominated experiments.

Various examples of majorized, non-majorized, weakly dominated and discrete experiments are given in Section 1.3. In particular, it is shown in Example 1.3 that the "discrete experiment" of Basu and Ghosh (1967) is a special case of a weakly dominated experiment.

In Chapter 2 we first construct the smallest PSS subfield for a majorized experiment. The construction is very intuitive and easy in the sense that the smallest PSS subfield is generated by all the likelihood ratios and supports. Then, using a maximal decomposition of the marginal experiment of the smallest PSS subfield, we show the existence of a pivotal measure. The pivotal measure then, in Section 2.2, is applied to prove generalizations of the Neyman factorization theorem.

In Section 2.3 we especially consider such generalizations and pivotal measures in the weakly dominated case. Then the Neyman factorization of the familiar type in the case of domination is exactly equivalent to the PSS property of the subfield. The smallest PSS is used to show the existence of the smallest sufficient subfield for weakly dominated experiments. The latter is simply the "weak" completion of the former. As for the pivotal measure we gradually recognize that it is something which plays a role to show "equivalence" of the original experiment to the marginal experiment of the smallest sufficient subfield. A precise explanation is given in Section 2.3.

The Halmos-Savage-Bahadur theory is treated, in Section 2.4, as a special case of the previous theory.

The first part of Chapter 3 deals with discrete experiments. Here a statistic is defined as a partition of the sample space. We show that a smallest sufficient statistic can be constructed just in the same way as Lehmann and Scheffé (1950) did in the case of domination under the assumption of "separability". But the Burkholder pathology may occur in the case of a discrete experiment. The detailed structure of pairwise sufficient and PSS subfields is given in Section 3.1. Both properties can be stated in terms of the smallest sufficient statistic. A subfield is pairwise sufficient if and only if it separates the partition of the smallest sufficient statistic. On the other hand a PSS subfield is characterized as the subfield which includes the partition of the smallest sufficient statistic. These observations clearly indicate the difference between pairwise sufficient and PSS subfield.

In Section 3.2, we proceed to assert that the properties of pairwise sufficiency and PSS stated in Section 3.1 are preserved in passing for majorized experiments. But in this more general situation the maximal decomposition of the experiment plays the same role as the smallest sufficient statistic does in the discrete case. Each part of the decomposition generates a dominated experiment by considering conditional probabilities. Then the smallest PSS subfield for the original experiment consists of sets which belong to the smallest sufficient subfield on each part of the decomposition, and which meet countably many parts of the decomposition or which are unions of all the parts but a countable number of parts.

In Section 3.3 we give a necessary and sufficient condition for a subfiled which contains a sufficient one to be sufficient. The condition is concerned with "closedness" of taking special type of completion of the subfield.

In Chapter 4 we study versions of the Rao-Blackwell theorem and a UMVUE(uniformly minimum variance unbiased estimator) theory for the undominated case. In Section 4.1 a PSS subfield is characterized as the subfield which has a property very similar to the Rao-Blackwell property. We show that a subfield is PSS if and only if, for any quadratically estimable parameter function, any unbiased estimator can be improved by an unbiased one which is measurable w.r.t. the subfield on each support.

Bahadur (1957) showed, in the dominated case, that if for every quadratically estimable parameter function there is a UMVUE then there exists a quadratically complete sufficient subfield. This is a converse of a theorem of Lehmann and Scheffé (1950). Torgersen(1988) extended Bahadur's result to majorized experiments. His result is presented in Section 4.2.

In Chapter 5 we construct conditional probability which is independent of the parameter, that is common conditional probability, in an extended sense for a given PSS subfield. This gives an extension of the notion of common conditional probability for a sufficient subfield. Technically we use the concrete representations of the L-space and M-space of an experiment. These notions from the theory of comparison of experiments were introduced by Le Cam (1964). In Section 5.1 we therefore collect some basic notions and results from this theory. We also construct the transition which renders the deficiency of the marginal experiment corresponding to the smallest PSS subfield w.r.t. the original experiment zero. In the definition of such a transition the pivotal measure plays an essential role as was conjectured in Chapter 2. This is due to the concrete representation of the L-space of a majorized experiment.

Section 5.2 is devoted to show the correspondence of the two representations of the M-space of a majorized experiment given by McShane (1962) and Torgersen (1979).

Using the Torgersen representation of the M-space an extended notion of common conditional probability for a PSS subfield is defined in Section 5.3. It is given as the dual mapping of the transition introduced in Section 5.1 and stated above. Hence the notion of pivotal measure again contributes considerably to the theory.

This extended notion of common conditional probability is applied in Section 5.4 to show Rao-Blackwell and Lehmann-Scheffé type theorems for PSS subfields.

In the final Chapter 6 we study the structure of pivotal measures using the L-spaces of experiments. After having introduced in Section 6.1 the important notion of a minimal L-space, we prove a characterization theorem for pivotal measures in Section 6.2. Any pivotal measure is given as the sum of elements of a maximal orthogonal system of the L-space of the majorized marginal experiment corresponding to the smallest PSS subfield. Conversely any sum of elements of the maximal orthogonal system is a pivotal measure. Taking a very special type of sum we regain the Halmos and Savage pivotal measure from which our discusssion started.

The bibliographical notes at the end of each Chapter provides the references, related topics not covered in the text.

Each Theorem, Lemma, Corollary etc. has serial numbers in a Chapter.

Chapter 1. Undominated experiments

In this chapter we briefly discuss two undominated classes of experiments. A dominated experiment has a dominating measure which is σ-finite. Hence if we drop the assumption "σ-finite" we have an undominated experiment. We call it majorized. At this stage of study this majorized experiment is recognized to be the most general experiment in which some "theory of sufficiency" could be constructed.

An intermediate concept between majorized experiment and dominated experiment is considered here and is called a weakly dominated experiment. A weakly dominated experiment has a localizable majorizing measure. Any σ-finite measure is localizable. The counting measure on the power set of an uncountable set is localizable and is not σ-finite. A statistical experiment which has such a counting measure as a majorizing measure is a model of sampling from a finite population, and is an important example of undominated experiments.

Final section of this chapter contains various examples of undominated experiments which fall into majorized or weakly dominated experiments.

Main references of this chapter are Diepenbrock(1971), Ghosh, Morimoto and Yamada (1981), and Mussmann (1972).

1.1. Majorized experiments and their decomposition

In this section we show that a majorized experiment is characterized by the existence of a support. This notion of support with another important notion of decomposition of experiment is applied to show that any majorized experiment has an equivalent majorizing measure.

An *experiment* $\mathscr{E}$ is a triplet $(\mathscr{X},\mathscr{A},\mathscr{P}=\{P_\theta;\theta\in\Theta\})$ consisting of a set $\mathscr{X}$, a σ-field $\mathscr{A}$ of subsets of $\mathscr{X}$ and a family $\mathscr{P}$ of probability measures on $(\mathscr{X},\mathscr{A})$. For any measure m on $(\mathscr{X},\mathscr{A})$ we denote the family of m-null sets by $N(m)=\{A\in\mathscr{A};m(A)=0\}$. Then for any $\mathscr{Q}\subset\mathscr{P}$ $N(\mathscr{Q})$ is defined by $N(\mathscr{Q})=\{A\in\mathscr{A};A\in N(P_\theta)$ for all $P_\theta\in\mathscr{Q}\}$. If $N(m)\subset N(\mathscr{P})$ then we write $\mathscr{P}\ll m$.

<u>Definition 1.1</u>. $\mathscr{E}$ is said to be *majorized* if there exists a measure m on $(\mathscr{X},\mathscr{A})$ such that each P_θ in $\mathscr{P}$ has a density dP_θ/dm w.r.t. m. Such an m is called a *majorizing*

measure for $\mathscr{E}$. If $N(\mathcal{P})=N(m)$, denoted simply by $\mathcal{P}\sim m$, then m is called an *equivalent majorizing measure* .

Definition 1.2. Let $\theta\in\Theta$. Then a set $S(\theta)\in\mathcal{A}$ having the following properties is called a *support* of P_θ for $\mathscr{E}$.

(1) $P_\theta(S(\theta)) = 1$, and

(2) $A\in\mathcal{A}$, $A\subset S(\theta)$ and $P_\theta(A) = 0$ imply $A\in N(\mathcal{P})$.

Definition 1.3. Let $\mathcal{B}$ be a subfield of $\mathcal{A}$. If there exists a support of P_θ belonging to $\mathcal{B}$ for all θ, then $\mathcal{B}$ is said to contain supports .

This definition of support is a slight extension of the usual definition, the positive part of the density $[dP_\theta/dm>0]$. Here for any propositional function $P(x)$ of x, $[P(x)]$ denotes the set of all x in $\mathcal{X}$ which satisfy $P(x)$. If m is equivalent to $\mathcal{P}$ and $\mathcal{B}$ contains supports, then by the uniqueness property of the support (that is, the symmetric differnce $S(\theta)\Delta S'(\theta)$ belongs to $N(\mathcal{P})$), there exists a version $p(x,\theta)$ of the density of P_θ w.r.t.m such that $[p(x,\theta)>0]\in\mathcal{B}$. Hence any support is σ-finite w.r.t. any equivalent majorizing measure. But if m is not equivalent to $\mathcal{P}$ then there exists an example in which the extended definition of support plays essential role (Chapter 2).

Next we introduce the notion of partition of an experiment. It is applied to construct an equivalent majorizing measure for a majorized experiment and to describe the structure of the smallest PSS. Let $\mathscr{E}=(\mathcal{X},\mathcal{A},\mathcal{P}=\{P_\theta;\theta\in\Theta\})$ be any majorized experiment. Let us fix any support of P_θ for each θ and denote it by $S(\theta)$. Then, the totality of decompositions of $\mathscr{E}$,

$\bar{\mathcal{F}}=\{\mathcal{F};$ (1) $\mathcal{F}\subset\mathcal{A}$,
(2) for any two different F_1,F_2 in $\mathcal{F}$ $F_1\cap F_2\in N(\mathcal{P})$, and (3) for any $F\in\mathcal{F}$ there exists a P_θ such that $P_\theta(F)>0$, $F-S(\theta)\in N(\mathcal{P})\}$,

is not empty and satisfies the condition of Zorn's Lemma w.r.t. the order of inclusion of the family of sets. Hence there exists a maximal element in $\bar{\mathcal{F}}$ which we denote by $\mathcal{F}=\{F\}$. We call $\mathcal{F}$ a *maximal decomposition* of $\mathscr{E}$. For each F in $\mathcal{F}$ the corresponding distribution in $\mathcal{P}$ conditioned by (3) is denoted by P_F. Clearly we can take F so as to satisfy $F\subset S(P_F)$ for each $F\in\mathcal{F}$. Hereafter we consider such a maximal decomposition. Let us define a set function n on $\mathcal{A}$ by

$$n(A)=\sum\{P_F(A\cap F)\,;F\in\mathcal{F}\}. \tag{1.1}$$

The sum means the supremum of

$$\{\sum P_F(A\cap F)\,;F\in\mathcal{F}_0\subset\mathcal{F} \text{ and } \mathcal{F}_0 \text{ is finite}\}.$$

To show that this n is an equivalent majorizing measure for $\mathcal{E}$ we need a few lemmas. In the following $\mathcal{A}_\sigma(n)$ shows the totality of sets with σ-finite measure w.r.t. n, and $\mathcal{A}(m)$ denotes the family of sets with finite measure w.r.t. m. Notation $m\ll n$ is used to indicate that m is absolutely continuous w.r.t. n. For any families $\mathfrak{m}_1$, $\mathfrak{m}_2$ of measures we write $\mathfrak{m}_1\ll\mathfrak{m}_2$ if $N(\mathfrak{m}_2)\subset N(\mathfrak{m}_1)$.

<u>Lemma 1.1</u>. Let m and n be measures on $(\mathcal{X},\mathcal{A})$ such that $m\ll n$ and let m be a finite measure. Then m has a density w.r.t. n if and only if there exists $T\in\mathcal{A}_\sigma(n)$ such that

$$m(A)=m(A\cap T)$$

for all $A\in\mathcal{A}$.

<u>Proof</u>. "If" part. There exists a density on T of m w.r.t. n, which we denote by f. Then the function

$$\begin{aligned} g(x) &= f(x) && \text{if } x\in T, \\ &= 0 && \text{if } x\in\overline{T}, \end{aligned}$$

is a density of m w.r.t. n.

Conversely suppose that m has the density $dm/dn(x)$ w.r.t. n. Then $T=[dm/dn(x)>0]$ satisfies the condition.

A measure m on $(\mathcal{X},\mathcal{A})$ is said to have the *finite subset property* if, for any $A\in\mathcal{A}$ such that $\infty\geq m(A)>0$, there exists $B\subset A$ such that $0<m(B)<\infty$.

<u>Lemma 1.2</u>. A measure m on $(\mathcal{X},\mathcal{A})$ has the finite subset

property if and only if any $A\in\mathcal{A}$ such that $m(A\cap E)=0$ for all $E\in\mathcal{A}(m)$ satisfies $m(A)=0$.

Proof. "Only if" part is a direct consequence of the definition. "If" part is also easy to show. Take any $A\in\mathcal{A}$ such that $\infty\geq m(A)>0$. Then by the assumption there exists an $E\in\mathcal{A}(m)$ such that

$$m(A\cap E)>0.$$

This clearly implies that m has the finite subset property.

For any measure m on $(\mathcal{X},\mathcal{A})$ and any $B\in\mathcal{A}$, we define a measure m/B on $(B,B\cap\mathcal{A})$ by $m/B(\cdot)=m(\cdot)$, $\cdot\in B\cap\mathcal{A}$. Here

$$B\cap\mathcal{A}=\{B\cap A;A\in\mathcal{A}\}.$$

Lemma 1.3. Let m and n be measures on $(\mathcal{X},\mathcal{A})$ such that $n\ll m$, and let n have the finite subset property. Then we have $\mathcal{A}_\sigma(m)\subset\mathcal{A}_\sigma(n)$.

Proof. Let us take any $B\in\mathcal{A}(m)$. By Lemma 1.2, n/B on $(B,B\cap\mathcal{A})$ has the finite subset property. Then there exist subsets B_1 and B_2 of B in $B\cap\mathcal{A}$, such that $B_1\cup B_2=B$, B is σ-finite w.r.t. n/B and B_2 satisfies $n/B(B_2\cap E)=0$ for any $E\in B\cap\mathcal{A}$ such that $n/B(E)<\infty$. This follows from exercise (11) of section 31 of Halmos(1950). Hence $B\in\mathcal{A}_\sigma(n)$.

Theorem 1.1. The measure defined by (1.1) is an equivalent majorizing measure for $\mathcal{E}$.

Proof. It is easy to show $n\ll\mathcal{P}$. To prove the converse let us take any $N\in\mathcal{A}$ such that $n(N)=0$. If there exists a $P_\theta\in\mathcal{P}$ such that $P_\theta(N)>0$, we have, for $A=N\cap S(\theta)$, $P_\theta(A)>0$ and $A - S(\theta) = \emptyset$ therefore for any $P\in\mathcal{P}$ and $F\in\mathcal{F}$,

$$P(A\cap F) = P(A\cap F\cap S(\theta))+P(A\cap F\cap(\mathcal{X}-S(\theta)))=0.$$

Hence $\mathcal{F}\cup\{A\}\in\bar{\mathcal{F}}$ and $A\bar{\in}\mathcal{F}$. This contradicts the maximality of $\mathcal{F}$ in $\bar{\mathcal{F}}$. Hence we have $\mathcal{P}\sim n$.

For any $A\in\mathcal{A}$ and $F\in\mathcal{F}$ we have $n(A\cap F)=P_F(A\cap F)$, so it follows that n has the finite subset property. Take $P_\theta\in\mathcal{P}$. Then

$$n/S(\theta) \ll P_\theta/S(\theta)$$

and $n/S(\theta)$ has the finite subset property. By Lemma 1.3

$$\mathcal{A}_\sigma(P_\theta/S(\theta)) \subset \mathcal{A}_\sigma(n/S(\theta)).$$

Hence, in particular, it follows $S(\theta) \in \mathcal{A}_\sigma(n)$. By Lemma 1.1, P_θ has a density w.r.t. n.

The following is the main result of this section.

Theorem 1.2. The following conditions are all equivalent.
(1) Each P_θ in $\mathcal{P}$ has a support for $\mathcal{E}$.
(2) $\mathcal{E}$ is majorized.
(3) There exists an equivalent majorizing measure for $\mathcal{E}$.

This follows from Theorem 1.1, so we omit the proof.

Next we show $\mathcal{F}$ is a maximal decomposition of the measure space $(\mathcal{X},\mathcal{A},n)$. Suppose that $(\mathcal{X},\mathcal{A},m)$ is a measure space with the finite subset property. Let us consider the family

$$\bar{\mathcal{D}} = \{\mathcal{D};\ (1)\ \mathcal{D}\subset\mathcal{A},\ (2)\ 0<m(A)<\infty \text{ for any } A\in\mathcal{D},\ (3) \text{ for any different } D_1 \text{ and } D_2\in\mathcal{D}, D_1\cap D_2\in N(m)\},$$

the totality of decompositions of the measure space. Clearly $\bar{\mathcal{D}}$ is not empty, because m has the finite subset property, and is an inductive set w.r.t. the order of inclusion. So $\bar{\mathcal{D}}$ has a maximal element by Zorn's Lemma. This maximal element is called a maximal decomposition of $(\mathcal{X},\mathcal{A},m)$.

Lemma 1.4. $\mathcal{D}(\in\bar{\mathcal{D}})$ is a maximal element of $\bar{\mathcal{D}}$ if and only if $m(A\cap D)=0$ for all $D\in\mathcal{D}$ imply that $m(A)=0$.

Proof is easy, so we omit it.

By Lemma 1.4 it follows that the maximal decomposition $\mathcal{F}$ of $\mathcal{E}$ is a maximal decomposition of $(\mathcal{X},\mathcal{A},n)$.

Proposition 1.1. Take any maximal decomposition $\mathcal{D}$ of $(\mathcal{X},\mathcal{A},m)$ and $A\in\mathcal{A}(m)$. Then $\{D\in\mathcal{D}; m(A\cap D)>0\}$ is countable, say $\{D_i; i\geq 1\}$, and

$$m(A - \bigvee_{i=1}^{\infty} D_i) = 0.$$

1.2. Weakly dominated experiments

A weakly dominated experiment is a majorized experiment for which there exists a localizable majorizing measure. In such a case there exists the smallest sufficient subfield, but there may exist the pathological example of Burkholder. Discrete experiment, which was introduced by Basu and Ghosh (1967) as a measure theoretical framework of the sampling from a finite population, constitutes an important example of weak domination. The following property of a weakly dominated experiment is itself important: for any experiment there exists a weakly dominated experiment such that the corresponding L- and M-spaces are isomorphic. But we note many important statistical properties such as the existence of a minimal sufficient subfield, etc., are not preserved by this "equivalency". We will not study this property. For this result we refer to Siebert(1979a), Luschgy and Mussmann(1985).

Before giving the formal definition of a weakly dominated experiment, we would like to show a few basic properties of localizable measures. A more detailed discussion about localizable measures can be found in Zaanen(1967).

Definition 1.4. A measure m on $(\mathcal{X},\mathcal{A})$ is said to be *localizable* if for any $\mathcal{G}\subset\mathcal{A}(m)$ there exists $A\in\mathcal{A}$, which is called an *essential supremum* of $\mathcal{G}$ w.r.t. m, with the following properties.

(1) $m(G - A)=0$ for all $G\in\mathcal{G}$,

and

(2) for any B in $\mathcal{A}$ such that $m(G-B)=0$ for all $G\in\mathcal{G}$ it follows that $m(A-B)=0$.

Proposition 1.2. If m is σ-finite then it is localizable.

Proof. Let $\mathcal{X}$ be the countable disjoint union of E_i with $0<m(E_i)<\infty$ for $i=1,2,\ldots$. Let us take any $\mathcal{G}\subset\mathcal{A}(m)$. Then $\{G\in\mathcal{G};$

$m(E_i \cap G)>0\}$ is countable, say,

$$G_1^i, G_2^i, \ldots .$$

Then it is easy to see that

$$G_0 = \bigcup_{i=1}^{\infty} \bigcup_{j=1}^{\infty} G_j^i$$

satisfies the condition of essential supremum of $\mathcal{G}$ w.r.t. m given in definition 1.4.

<u>Definition 1.5</u>. A family of $\mathcal{A}$-measurable functions

$$\{f(x,E); E \in \mathcal{A}(m)\}$$

satisfying $f(x,E)=0$ outside E, and

$$I_{E_1 \cap E_2} f(x,E_1) = I_{E_1 \cap E_2} f(x,E_2)$$

a.e.m for all $E_1, E_2 \in \mathcal{A}(m)$, is called a *cross-section* w.r.t. m. Here I_E denotes the indicator function of E and "a.e.m" means "almost everywhere w.r.t. m".

<u>Theorem 1.3</u>. Suppose that m has the finite subset property. Then in order that m is localizable it is necessary and sufficient that for any cross-section $\{f(x,E); E \in \mathcal{A}(m)\}$ there exists an $\mathcal{A}$-measurable function f such that $f(x) I_E(x) = f(x,E)$ a.e.m for all $E \in \mathcal{A}(m)$.

For a proof we refer to Zaanen (1967, section 35, Theorem 2).

Now we give the definition of a weakly dominated experiment.

<u>Definition 1.6</u>. An experiment $\mathcal{E}=(\mathcal{X},\mathcal{A},P=\{P_\theta;\theta\in\Theta\})$ is said to be *weakly dominated* if it is majorized by a localizable measure.

By Proposition 1.2 any majorized experiment with σ-finite dominating measure is weakly dominated. Such an experiment is simply called *dominated*.

For sets A and B we write $A\subset B\ [m]$ if $m(A-B)=0$.

Theorem 1.4. Let $\mathcal{E}=(\mathcal{X},\mathcal{A},\mathcal{P}=\{P_\theta;\theta\in\Theta\})$ be weakly dominated. Then it has an equivalent localizable majorizing measure.

Proof. Suppose that m is a localizable majorizing measure. Let us define $T(\theta)=[dP_\theta/dm(x)>0]$ and

$$\mathcal{F}=\{E\in\mathcal{A}(m);E\subset T(\theta)\ \text{for some}\theta\}.$$

Then, by assumption, there exists an essential supremum T of $\mathcal{F}$ w.r.t. m. Define n by $n(A)=m(A\cap T)$, $A\in\mathcal{A}$. It is easy to derive localizability of n from that of m.

Next we show that n has the finite subset property. Assume that $0<n(A)=m(A\cap T)$. Then $m(A\cap T(\theta))>0$ for some θ. Since $T(\theta)$ is σ-finite w.r.t. m there exists $E(\theta)\subset T(\theta)$ such that

$$0<m(A\cap E(\theta))\leq m(E(\theta))<\infty.$$

Since $E(\theta)\subset T(\theta)\subset T\ [m]$, n has the finite subset property.

By Lemma 1.2 each $T(\theta)$ belongs to $\mathcal{A}_\sigma(n)$. Hence P_θ has a density w.r.t. n by Lemma 1.1.

Suppose that $A\in N(\mathcal{P})$. Because $m(A\cap T(\theta))=0$ for all $\theta\in\Theta$, we have $m(E-(\mathcal{X}-A))=m(E\cap A)=0$ for all $E\in\mathcal{F}$. By the definition of T it follows that $m(T\cap A)=m(T-(\mathcal{X}-A))=0$, which implies $n(A)=0$.

Definition 1.7. (1) A family of real valued $\mathcal{A}$-measurable functions $\{f(x,\theta);\theta\in\Theta\}$ is said to be *essentially bounded* if there exists a constant M such that

$$sup\{ess.sup_{P_\theta}|f(x,\theta)|;\theta\in\Theta\}\leq M<\infty.$$

(2) An essentially bounded family $\{f(x,\theta);\theta\in\Theta\}$ is called *finitely* (resp. *countably*) *coherent* if, for any finite (resp. countable) subset Θ_0 of Θ, there exists an $\mathcal{A}$-measurable function f such that $f(x)=f(x,\theta)$ a.e. P_θ for all $\theta\in\Theta_0$.

Remark 1.1. Any finitely coherent family is countably coherent. This follows by extending the sequence of the "coherent functions".

Theorem 1.5. An experiment $\mathcal{E}=(\mathcal{X},\mathcal{A},\mathcal{P}=\{P_\theta;\theta\in\Theta\})$ is weakly dominated if and only if for any essentially bounded finitely coherent family $\{f(x,\theta);\theta\in\Theta\}$ there exists an $\mathcal{A}$-measurable function f such that $f(x)=f(x,\theta)$ a.e. P_θ for all $\theta\in\Theta$.

Proof. Let m be an equivalent localizable majorizing measure for $\mathcal{E}$. Then for any subfamily $\mathcal{C}\subset\mathcal{A}$ there exists a set $A\in\mathcal{A}$, which is called a simultaneous essential supremum, such that, for all $\theta\in\Theta$,

(1) $P_\theta(C-A)=0$ for all $C\in\mathcal{C}$,

(2) if $B\in\mathcal{A}$ satisfies that $P_\theta(E-B)=0$ for all $E\in\mathcal{C}$ then $P_\theta(A-B)=0$.

To prove this let us define $\mathcal{C}_0=\mathcal{C}\cap\mathcal{A}(m)$ and let $A\in\mathcal{A}$ be the essential supremum of $\mathcal{C}_0$ w.r.t. m, whose existence is guaranteed by the localizability of m. This A satisfies the required conditions of (1) and (2) above.

The concept of simultaneous supremum can be extended for any family of $\mathcal{A}$-measurable functions. Let f be the simultaneous supremum of the family

$$\{I_{\mathcal{X}-A(\theta)}(x)f(x,\theta);\theta\in\Theta\},$$

where $A(\theta)$ is the simultaneous supremum of $\{A\in\mathcal{A};P_\theta(A)=0\}$. Then, using the finitely coherent property of $\{f(x,\theta);\theta\in\Theta\}$, it can be shown that this f satisfies $f(x)=f(x,\theta)$ a.e. P_θ for all $\theta\in\Theta$.

Conversely suppose that any essentially bounded finitely coherent family $\{f(x,\theta);\theta\in\Theta\}$ has an $\mathcal{A}$-measurable function f such that $f(x)=f(x,\theta)$ a.e. P_θ for all $\theta\in\Theta$. First of all notice that there exists a support $T(\theta)$ of P_θ for $\mathcal{E}$ for each $\theta\in\Theta$. To see this, from the Lebesgue decomposition of P_θ w.r.t. P_{θ_0}, there exists a set T_{θ,θ_0} such that $P_\theta(T_{\theta,\theta_0})=1$ and $P_{\theta_0}\ll P_\theta$ on T_{θ,θ_0}. Then

$$\{I_{T_{\theta,\theta_0}}(x);\theta_0\in\Theta\}$$

is an essentially bounded finitely coherent family. Hence, by the assumption, there exists an $\mathcal{A}$-measurable function f_θ such that

$$f_\theta(x) = I_{T_{\theta,\theta_0}}(x) \text{ a.e. } P_{\theta_0}$$

for all $\theta_0 \in \Theta$. Then $T_\theta = f_\theta^{-1}(\{1\})$ satisfies the conditions of support of P_θ. Suppose that $\mathcal{F}_0$ is a maximal decomposition for $\mathcal{E}$. Define a set function n on $(\mathcal{X},\mathcal{A})$ by

$$n(A) = \sum\{P_F(A \cap F); F \in \mathcal{F}_0\}, \quad A \in \mathcal{A},$$

in the same way as (1.1). Then it can be proved that n is an equivalent localizable majorizing measure for $\mathcal{E}$, similarly to the proof of Theorem 1.1.

1.3. Examples

In this section we introduce a few examples. In example 1.1 the famous Pitcher's(1957) and Burkholder's(1961) examples are presented under a generalized treatment of uniform distributions. In the second example we introduce an experiment in which pairwise sufficiency does not necessarily imply PSS or sufficiency. The third example introduces the discrete experiment. And final example treats non-majorized experiments.

Example 1.1. Let $(\mathcal{X},\mathcal{A},m)$ be a measure space in which m has the finite subset property. Take any non-empty subfamily $\mathcal{F} = \{F(\theta); \theta \in \Theta\}$ of $\mathcal{A}(m)$ and suppose $m(F(\theta)) > 0$ for all $\theta \in \Theta$. Consider a statistical experiment $\mathcal{E} = (\mathcal{X},\mathcal{A},\mathcal{P} = \{P_\theta; \theta \in \Theta\})$, where

$$P_\theta(A) = m(A \cap F(\theta)) / m(F(\theta))$$

for all $A \in \mathcal{A}$. This is a majorized experiment, because $F(\theta)$ is a support of P_θ for $\mathcal{E}$. Pitcher's pathological example may occur in a special case of this type of experiment. Suppose that $\mathcal{X} = R^1$ and $\mathcal{A}$ = Borel field. Let S be a non-empty, non-Borel subset of $\mathcal{X}$ which does not contain 0 and is symmetric about the origin. Let m be the counting measure on $(\mathcal{X},\mathcal{A})$. For any $\theta \in \Theta = \mathcal{X}$ we define $F(\theta) = \{\theta, -\theta\}$ if $\theta \in S$, $F(\theta) = \{\theta\}$ if $\theta \in \bar{S}$. Then, by the definition of P_θ, it follows that $P_\theta(\theta) = P_\theta(-\theta) = 1/2$ if $\theta \in S$ and $P_\theta(\theta) = 1$ if $\theta \in \bar{S}$. For this experiment there does not exist the smallest sufficient subfield. We note that this experiment is not weakly dominated.

Another famous pathological phenomena discovered by Burkholder(1961), which is sometimes refered to as "Burkholder's phenomena", also occurs in this model. Suppose that $\mathcal{X}=R^1$ and $\mathcal{A}$ = Borel field. Let us define S and m in the same way as in the above example. For any $\theta\in\Theta=R^1$ we define $F(\theta)=\{\theta,-\theta\}$ if $\theta\neq 0$, $F(\theta)=\{0\}$ if $\theta=0$. Then P_θ satisfies $P_\theta(\theta)=P_\theta(-\theta)=1/2$ for $\theta\neq 0$, and $P_0(0)=1$. Then the subfield $\mathcal{B}$ of symmetric Borel sets is sufficient for this experiment. The subfield $\mathcal{C}$ which is defined by $\{A\in\mathcal{A};\ A\cap S$ is symmetric about the origin$\}$ is not sufficient though it includes $\mathcal{B}$.

<u>Example 1.2</u>. Let $\mathcal{X}=[0,1]\times\{0,1\}$, $\mathcal{A}$ = Borel field, and $\mathcal{P}=\{P_\theta;\theta\in[0,1]\}$. For each θ, P_θ is defined by

$$P_\theta((A\times\{0\})\cup(B\times\{1\}))=1/2(l(A)+I_B(\theta)),$$

where l is the Lebesgue measure on $[0,1]$. The subfield

$$\mathcal{B}=\{(A\times\{0\})\cup(A\times\{1\});A(\subset[0,1])\text{ is a Borel set}\}$$

is pairwise sufficient, but is not sufficient. This experiment is majorized, because

$$([0,1]\times\{0\})\cup(\{\theta\}\times\{1\})$$

is a support of P_θ. Note that $\mathcal{B}$ is not PSS. This example is due to Halmos and Savage(1949).

<u>Example 1.3</u>. Let us consider a finite population $\{1,2,\ldots,N\}$ and suppose that each member i has a real valued characteristic θ_i. For example θ_i is the height or weight of the i-th person. Then we define

$$\theta = (\theta_1,\theta_2,\ldots,\theta_N)\in R^N$$

as the parameter. Suppose $\Theta = R^N$ is the parameter space. Let us adopt the simple random sampling of size n without replacement. Then our data, x, consists of observations of the characteristics $(\theta_{i_1},\theta_{i_2},\ldots,\theta_{i_n})$. This means $i_1,i_2,\ldots,i_n$-th individuals were selected and their characteristics θ_{i_1}, $\theta_{i_2},\ldots,\theta_{i_n}$ were observed. We assume that there is no measuremennt error about the observation of the characteristic. Then P_θ, the distribution of the observation, is a probability measure on

$$(\mathcal{X}=R^n, 2^{\mathcal{X}}).$$

The support of P_θ is a finite subset of $\mathcal{X}$, so P_θ is a discrete probability measure. The statistical experiment thus obtained is not dominated. If this is dominated any equivalent dominating probability measure has a positive mass at each point of the sample space. Since the sample space is uncountable this is a contradiction.

We restricted to the case of a continuous quantity of observations. So the statistical experiment defined above, though a typical case, does not cover all the situations of sampling from a finite population. We now describe the experiment formally in the following way. An experiment $\mathcal{E}=(\mathcal{X},\mathcal{A},\mathcal{P}=\{P_\theta;\theta\in\Theta\})$ is said to be *discrete* if the following conditions are satisfied.

(1) $\mathcal{X}$ is an uncountable set,
(2) $\mathcal{A}$ is the power set of $\mathcal{X}$,
(3) each P_θ is discrete, and
(4) empty set is the only $\mathcal{P}$-null set.

Clearly discrete experiment is weakly dominated, because the counting measure on $(\mathcal{X},\mathcal{A})$ is an equivalent localizable majorizing measure for $\mathcal{E}$. This discrete experiment was introduced by Basu and Ghosh(1967). We will study discrete experiment in a later chapter.

Example 1.4. Here we give examples of non-majorized experiments. The first example is taken from Siebert(1979a). Let $\mathcal{X}=[0,2)$ and $\mathcal{A}$ = Borel field. For each $\theta\in[0,1)$ let

$$P_\theta = 1/2(\varepsilon_\theta+\varepsilon_{\theta+1}),$$

where ε_θ is the point probability measure at θ. P_1 and P_2 are defined as the Lebesgue measures on $[0,1)$ and $[1,2)$ respectively. Then the experiment $\mathcal{E}=(\mathcal{X},\mathcal{A},\mathcal{P}=\{P_\theta;\theta\in[0,1]\cup\{2\}\})$ is not majorized, simply because P_1 and P_2 do not have supports for $\mathcal{E}$.

Next example shows that non-majorized experiments are not restricted to this type of "mixture" of discrete distributions and continuous distributions. Let $\mathcal{X} = R^n, \mathcal{A} =$ Borel field, and let $\mathcal{P}$ be the totality of the continuous probability measures, or let $\mathcal{P}$ satisfy the condition that for each uncountable $A \in \mathcal{A}$ there exists $P \in \mathcal{P}$ such that $P(A)=1$. Then these experiments are not majorized. This was proved by Rao and Yamada(1980).

To prove the first case suppose that $\mathcal{P}$ is the totality of the continuous probability measures. Suppose that there exists a majorizing measure m for $\mathcal{E}$. Then, by a theorem of Kuratowski (1966, p.451), for any uncountable Borel subset B of $\mathcal{X}$ there exists a Borel isomorphism f from B to $\mathcal{X}$. Hence $f(B\mathcal{A})=f(\mathcal{A})$. Let us take a normal distribution n on $(\mathcal{X},\mathcal{A})$ and define $Q(A)=n(f(A))$ for $A \in B\mathcal{A}$. For any $A \in \mathcal{A}$ we put $P(A)=Q(A \cap B)$. Then it follows that $P(B)=1$ and $P(\{x\})=0$ for all $x \in \mathcal{X}$. Hence $P \in \mathcal{P}$. So we have $m(B)>0$. Since there exists an uncountable disjoint family of uncountable subsets of B, m is not σ-finite on B. Suppose P has a density g w.r.t. m. Then $[g(x)>0]$ is σ-finite w.r.t. m. Because of continuity of P, $[g(x)>0]$ is uncountable. Hence m is not σ-finite on $[g(x)>0]$. This is a contradiction.

1.4. Bibliographical notes

The nomenclature "majorized" is due to Siebert (1979a). The extended notion of support was introduced by Ghosh, Morimoto and Yamada(1981). The maximal decomposition of an experiment was defined for the first time by Diepenbrock(1971) to show equivalence between weak domination and Pitcher's compactness (1965), though Diepenbrock's definition of a localizable measure was slightly different from the present definition. In particular Theorem 1.5 is due to Diepenbrock. Pitcher's paper (1965) was a pioneering one which settled the framework of "compactness", in the undominated case. When Pitcher's(1957) and Burkholder's(1961) famous pathological examples were found there was no such a perspective. Mussmann (1972) introduced the nomenclature "weakly dominated", and the proof of Theorem 1.4 is due to him.

A decomposable experiment is a direct extension of a dominated experiment. An experiment $\mathcal{E}=(\mathcal{X},\mathcal{A},\mathcal{P})$ is *decomposable* if

(1) there exists a partition $(A_i)_{i \in I}$ of $\mathcal{X}$ such that a

set $A \subset \mathcal{X}$ belongs to $\mathcal{A}$ if and only if $A \cap A_i \in \mathcal{A}$ for all $i \in I$, and

(2) There exists a majorizing measure, called a strictly localizable measure, such that $m(A_i) < \infty$ for all i and $m(A) = \sum_i m(A \cap A_i)$.

Torgersen(1991) calls it a $\sum$-*dominated* experiment. A decomposable experiment is weakly dominated because any strictly localizable measure is localizable. Any dominated or discrete experiment is decomposable. We quoted in Section 1.2 that any experiment is equivalent to a weakly dominated experiment. Actually Luschgy and Mussmann (1985) proved that any experiment is equivalent to a weakly dominated experiment whose "completion" is decomposable.

In Example 1.3 we disregarded the part of "label" from the data. Lehmann (1983,Chapter 3,Section 6) gave a discussion about discarding the labels when they are unrelated to the associated values. Cassel,Sarndal and Wretman(1977,Section 2.4,Remark 3),Basu (Ghosh(ed.)(1988) Chapter 12,Section 4) also consider the same problem.

Luschgy and Mussmann(1987) characterized weak domination by the property of compactness of a topology on the set of (randomized) decision functions.

Pitcher(1965) raised the open problem whether the union or product of two weakly dominated experiments is weakly dominated. As for the union we refer to Ramamoorthi and Yamada (1983), and to Kusama and Yamada(1972), Morimoto(1972) for product. There exist counter examples for both union and product. Luschgy and Mussmann(1986) showed any finite number of products of majorized experiments are majorized. A detailed exposition of the notion of compactness by Pitcher(1965) was conducted by Morimoto(1973).

Under the condition that each uncountable separable metric space contains a non-Borel set, Rogge(1972) showed any weakly dominated experiment $(\mathcal{X}, \mathcal{A}, \mathcal{P})$ in which $\mathcal{A}$ is countably generated is dominated. Rogge, among others, actually showed this condition is also necessary for such weakly dominated experiments to be dominated. For such a set-theoretical aspect of discussion related to weakly dominated experiments we refer to Diepenbrock(1971), Kusama and Yamada(1972), Morimoto(1972), and Luschgy and Mussamnn(1985).

The book by Torgersen(1991), which is very comprehensive and excellent, also contains many results in undominated cases. But the book and this note have a few in common.

Chapter 2. PSS, pivotal measure and Neyman factorization

In this chapter we introduce PSS subfield. We also define and construct a special majorizing measure called a pivotal measure. Among PSS subfields there exists the smallest one, and the pathological phenomena found by Burkholder(1961) do not occur. A PSS subfield is characterized by a Neyman factorization. Pivotal measure plays a central role to prove these results.

Research on this chapter was done mainly by Ghosh, Morimoto and Yamada(1981), Ramamoorthi and Yamada(1982), and Yamada(1982).

2.1. PSS and pivotal measure for majorized experiments

Let $\mathcal{E}=(\mathcal{X},\mathcal{A},\mathcal{P}=\{P_\theta;\theta\in\Theta\})$ be any experiment. For any subfields $\mathcal{B}$ and $\mathcal{C}$, $\mathcal{B}\subset\mathcal{C}[\mathcal{P}]$ means that for any $B\in\mathcal{B}$ there exists $C\in\mathcal{C}$ such that $B\Delta C\in N(\mathcal{P})$.

Definition 2.1. A subfield $\mathcal{B}$ is said to be *sufficient* for $\mathcal{E}$ if, for any $A\in\mathcal{A}$, there exists a $\mathcal{B}$-measurable function $E(I_A/\mathcal{B})$, common conditional probability of A given $\mathcal{B}$, such that

$$P_\theta(A\cap B)=\int_B E(I_A/\mathcal{B})(x)dP_\theta \quad \text{for all } B\in\mathcal{B} \text{ and } \theta\in\Theta.$$

Hence $E(I_A/\mathcal{B})$ is the conditional probability of A given $\mathcal{B}$ w.r.t. P_θ for all θ, that is

$$E(I_A/\mathcal{B})(x)=E(I_A/\mathcal{B},P_\theta)(x) \quad \text{a.e. } P_\theta$$

for all θ. A sufficient subfield $\mathcal{B}$ is said to be the *smallest* if $\mathcal{B}\subset\mathcal{C}[\mathcal{P}]$ for any other sufficient subfield $\mathcal{C}$. If $\mathcal{B}$ is sufficient for any pair from $\mathcal{P}$ then $\mathcal{B}$ is called *pairwise sufficient* for $\mathcal{E}$.

Definition 2.2. Suppose that $\mathcal{E}$ is a majorized experiment. A subfield $\mathcal{B}$ is said to be *PSS*(pairwise sufficiency with supports) for $\mathcal{E}$ if it is pairwise sufficient and there exists a support $S(\theta)$ of P_θ belonging to $\mathcal{B}$ for each $\theta\in\Theta$. The term "PSS" is sometimes used as an adjective, i.e. PSS subfield (pairwise sufficient subfield with supports). A PSS subfield is called the *smallest* if it holds that $\mathcal{B}\subset\mathcal{C}$ $[\mathcal{P}]$ for any other PSS $\mathcal{C}$.

We define the strong completion of a subfield $\mathcal{C}$ by

$$\bar{\mathcal{C}}(\mathcal{A})=\{A\in\mathcal{A};\ \text{there exists } C\in\mathcal{C} \text{ such that } A\Delta C\in N(P_\theta) \text{ for all } \theta\in\Theta\}.$$

Then $\mathcal{B}$ is the smallest PSS if and only if $\mathcal{B}\subset\bar{\mathcal{C}}(\mathcal{A})$ for any other PSS $\mathcal{C}$.

In the next Lemma we use the notation $E(f/\mathcal{B},m)$, which denotes the conditional expectation of f given $\mathcal{B}$ w.r.t. m. Here f is an $\mathcal{A}$-measurable and m-integrable function.

Lemma 2.1. Let $\mathcal{E}=(\mathcal{X},\mathcal{A},\mathcal{P}=\{P_\theta;\theta\in\Theta\})$ be any experiment and let $\mathcal{B}$ be a subfield of $\mathcal{A}$.

(1) $\mathcal{B}$ is pairwise sufficient if and only if, for any pair $\{P_{\theta_1},P_{\theta_2}\}$, there exists a $\mathcal{B}$-measurable density of P_{θ_1} w.r.t. $P_{\theta_1}+P_{\theta_2}$.

(2) $\mathcal{B}$ is pairwise sufficient if and only if, for any countable subfamily $\mathcal{P}_0$ of $\mathcal{P}$, $\mathcal{B}$ is sufficient for $\mathcal{P}_0$.

Proof. (1) Let f be a $\mathcal{B}$-measurable density of P_{θ_1} w.r.t. $P_{\theta_1}+P_{\theta_2}$. Then for any $A\in\mathcal{A}$,

$$E(I_A/\mathcal{B},P_{\theta_1})(x)f(x)=E(I_A/\mathcal{B},P_{\theta_1}+P_{\theta_2})(x)f(x) \quad \text{a.e. } P_{\theta_1},$$

and

$$E(I_A/\mathcal{B},P_{\theta_2})(x)f(x)=E(I_A/\mathcal{B},P_{\theta_1}+P_{\theta_2})(x)f(x) \quad \text{a.e. } P_{\theta_2}.$$

Hence $\mathcal{B}$ is sufficient for $\{P_{\theta_1},P_{\theta_2}\}$. Conversely take any θ_1, $\theta_2\in\Theta$. Then $E(\{dP_{\theta_1}/d(P_{\theta_1}+P_{\theta_2})\}/\mathcal{B},P_{\theta_1}+P_{\theta_2})$ is a $\mathcal{B}$-measurable density of P_{θ_1} w.r.t. $P_{\theta_1}+P_{\theta_2}$.

(2) Take any at most countable subfamily $\mathcal{P}_0=\{P_{\theta_1},P_{\theta_2},\ldots\}$ of $\mathcal{P}$ and put

$$m(A)=\sum_{i=1}^{\infty} 2^{-i}P_{\theta_i}(A), \quad A\in\mathcal{A}.$$

Then by Lemma 12 of Halmos and Savage(1949) there exists a $\mathcal{B}$-measurable density f_i of P_{θ_i} w.r.t. m, because $\mathcal{B}$ is pairwise sufficient for $\mathcal{P}$. Then we can easily prove that

$$E(I_A/\mathcal{B},m)(x)f_i(x)=E(I_A/\mathcal{B},P_{\theta_i})(x)f_i(x) \text{ a.e.} P_{\theta_i}$$

for all $i\geq 1$, hence $E(I_A/\mathcal{B},m)(x)=E(I_A/\mathcal{B},P_{\theta_i})(x)$ a.e.P_{θ_i} for all $i\geq 1$. Therefore $\mathcal{B}$ is sufficient for $\mathcal{P}_0$. The converse is clear.

Theorem 2.1. Let $\mathcal{E}=(\mathcal{X},\mathcal{A},\mathcal{P}=\{P_\theta;\theta\in\Theta\})$ be any majorized experiment. Then there exists the smallest PSS for $\mathcal{E}$.

Proof. Let $\mathcal{B}_0$ be the subfield which is generated by

$$\{dP_{\theta_1}/d(P_{\theta_1}+P_{\theta_2})I_{S(\theta_1)\cup S(\theta_2)},I_{S(\theta)};\theta,\theta_1,\theta_2\in\Theta\}.$$

By Lemma 2.1 $\mathcal{B}_0$ is PSS. Take any $\mathcal{B}$ which is PSS for $\mathcal{E}$. For any θ let $T(\theta)$ be a support of P_θ belonging to $\mathcal{B}$. We have $S(\theta)\Delta T(\theta)\in N(\mathcal{P})$ for all $\theta\in\Theta$. Suppose that $s(.;\theta_1,\theta_2)$ is a $\mathcal{B}$-measurable version of the density of P_{θ_1} w.r.t. $P_{\theta_1}+P_{\theta_2}$. Then, for any real number a, we have

$$[dP_{\theta_1}/d(P_{\theta_1}+P_{\theta_2})I_{S(\theta_1)\cup S(\theta_2)}>a]\Delta[s(x;\theta_1,\theta_2)I_{T(\theta_1)\cup T(\theta_2)}>a]$$

$\in N(\mathcal{P})$. This implies $\mathcal{B}_0\subset\mathcal{B}[\mathcal{P}]$.

Definition 2.3. An equivalent majorizing measure n for a majorized experiment $\mathcal{E}=(\mathcal{X},\mathcal{A},\mathcal{P}=\{P_\theta;\theta\in\Theta\})$ is called a *pivotal measure* for $\mathcal{E}$ if the following condition is satisfied; for any subfield $\mathcal{B}$ of $\mathcal{A}$, $\mathcal{B}$ is PSS if and only if there exists a $\mathcal{B}$-measurable density of P_θ w.r.t. n for each $\theta\in\Theta$.

Let m be a measure on $(\mathcal{X},\mathcal{A})$. Then, for a subfield $\mathcal{B}$ of $\mathcal{A}$, $m/\mathcal{B}$ denotes the restriction of m to $\mathcal{B}$.

Theorem 2.2. For any majorized experiment there exists a pivotal measure.

Proof. Let $\mathcal{B}_0$ be the smallest PSS for majorized experiment $\mathcal{E}=(\mathcal{X},\mathcal{A},\mathcal{P}=\{P_\theta;\theta\in\Theta\})$. Let $S(\theta)$ be a support of P_θ belonging to $\mathcal{B}_0$. Then the marginal experiment $\mathcal{E}(\mathcal{B}_0)=(\mathcal{X},\mathcal{B}_0,\mathcal{P}/\mathcal{B}_0=\{P_\theta/\mathcal{B}_0;\theta\in\Theta\})$ is majorized, because $S(\theta)$ is clearly a support of $P_\theta/\mathcal{B}_0$ for $\mathcal{E}/\mathcal{B}_0$. Let n be an equivalent majorizing measure for $\mathcal{E}/\mathcal{B}_0$ constructed by (1.1) through a maximal decomposition $\mathcal{F}=\{F\}$ of $\mathcal{E}/\mathcal{B}_0$,

$$n(B)=\sum\{P_F(B\cap F);F\in\mathcal{F}\},\ B\in\mathcal{B}_0. \quad (2.1)$$

As was shown in section 1.1, $\mathcal{F}$ is a maximal decomposition of $(\mathcal{X},\mathcal{B}_0,n)$.

For each $\theta\in\Theta$ $T(\theta)=[dP_\theta/\mathcal{B}_0/dn(x)>0]$ is σ-finite w.r.t. n. Hence there exists a countable subfamily $\{F_i;i\geq1\}$ of $\mathcal{F}$ satisfying

$$n(T(\theta)-\bigcup_{i\geq1}^{\infty}(T(\theta)\cap F_i))=0$$

by Proposition 1.1. Then it holds that $P_\theta(\bigcup_{i=1}^{\infty}F_i)=1$. Now let us use the same notation n as in (2.1) for the measure on $(\mathcal{X},\mathcal{A})$ defined by

$$n(A)=\sum\{P_F(A\cap F);F\in\mathcal{F}\},A\in\mathcal{A}. \quad (2.2)$$

Note that P_F in (2.1) is the restriction of P_F in (2.2) to $\mathcal{B}_0$, though they are denoted by the same notation.

We show that n in (2.2) is a pivotal measure for $\mathcal{E}$. Suppose that $n(A)=0$. Then $P_F(A\cap F)=0$ for any $F\in\mathcal{F}$. Take any $\theta\in\Theta$. Then from the above remark there exists a countable subfamily $\mathcal{F}_0=\{F_1,F_2,\ldots\}$ such that $P_\theta(\bigcup_{i=1}^{\infty}F_i)=1$. Hence we have

$$P_\theta(A) = \sum_{i=1}^{\infty}P_\theta(A\cap F_i)= 0,$$

from which it follows that $\mathcal{P}\sim n$.

Next we show each P_θ has a $\mathcal{B}_0$-measurable density w.r.t. n. For each $i\geq1$ let f_i be a $\mathcal{B}_0$-measurable version of the density of P_θ w.r.t. $P_\theta+P_{\theta_i}$. Let g_i be the function defined by

$$\begin{aligned} g_i(x)&=f_i(x), && \text{if } x\in\mathcal{X}-F_i\cap[f_i(x)=1],\\ &=1/2, && \text{if } x\in F_i\cap[f_i(x)=1]. \end{aligned}$$

Then

$$\begin{aligned} f(x)&=g_i(x)/(1-g_i(x)), && \text{if } x\in S(\theta)\cap[F_i-\bigcup_{j=1}^{i-1}F_j],\ i=1,2,\ldots,\\ &=0 && \text{otherwise}, \end{aligned}$$

is a $\mathcal{B}_0$-measurable density of P_θ w.r.t. n. Hence, for any PSS

$\mathcal{B}$, each P_θ has a $\mathcal{B}$-measurable density w.r.t. n.

Conversely if each P_θ has a $\mathcal{B}$-measurable version of the density w.r.t. n, then $\mathcal{B}$ contains supports, and is pairwise sufficient by Lemma 2.1.

2.2. Generalizations of the Neyman factorization theorem

Next theorem characterizes PSS by a Neyman factorization. Let $\mathcal{E}=(\mathcal{X},\mathcal{A},\mathcal{P}=\{P_\theta;\theta\in\Theta\})$ be a majorized experiment and let m be any majorizing measure for $\mathcal{E}$.

Theorem 2.3. A subfield $\mathcal{B}$ is PSS if and only if, for each $A\in\mathcal{A}_\sigma(m)$ and $\theta\in\Theta$, there exist a non-negative $\mathcal{B}$-measurable function $g(.,\theta)$ independent of A and an $\mathcal{A}$-measurable function $h(.,A)$ independent of θ such that

$$dP_\theta/dm(x)=g(x,\theta)h(x,A) \text{ a.e. } m \text{ on } A. \tag{2.3}$$

Proof. Let $\mathcal{B}$ be PSS and let n be any pivotal measure for $\mathcal{E}$. Then each P_θ has a $\mathcal{B}$-measurable density $g(.,\theta)$ w.r.t. n. Take any $A\in\mathcal{A}_\sigma(m)$. By Lemma 1.3, $A\in\mathcal{A}_\sigma(n)$. Let $h(.,A)$ be the function defined by

$$\begin{aligned} h(x,A)&=dn/dm(x) && \text{if } x\in A,\\ &=0 && \text{if } x\in\bar{A}.\end{aligned}$$

Then we have (2.3).

Conversely suppose that (2.3) holds. We first show $B_\theta=[g(x,\theta)>0]$ is a support of P_θ belonging to $\mathcal{B}$. Since $[dP_\theta/dm(x)>0]\in\mathcal{A}_\sigma(m)$, we have from (2.3) $P_\theta(B_\theta)=1$. Take $B\in\mathcal{A}$ such that $B\subset B_\theta$ and $P_\theta(B)=0$. Then by

$$0=P_\theta(A\cap B)=\int_{A\cap B\cap[h(x,A)>0]} g(x,\theta)h(x,A)\ dm\ ,\quad A\in\mathcal{A}_\sigma(m),$$

we have $m(A\cap B\cap[h(x,A)>0])=0$. Hence it follows that, for any $Q\in\mathcal{P}$,

$$Q([dQ/dm(x)>0]\cap B\cap[h(x,[dQ/dm(x)>0])>0])=0.$$

On the otherhand $Q([h(x,[dQ/dm(x)>0])=0)=0$ holds. So we have $Q(B)=0$. By the definition of a support, B_θ is a support of P_θ

belonging to $\mathcal{B}$.

Take any $\theta_1, \theta_2 \in \Theta$ and define a $\mathcal{B}$-measurable function $k(.)$ by

$$k(x) = g(x,\theta_1)/\{g(x,\theta_1)+g(x,\theta_2)\} \quad \text{if } x \in [g(x,\theta_1)+g(x,\theta_2)>0],$$
$$=0 \quad \text{otherwise.}$$

Then we have $(P_{\theta_1}+P_{\theta_2})((A-C)\cap B)=0$, where $A\in\mathcal{A}, B\in\mathcal{A}_\sigma(m)$,

$$C=A\cap(T_1\cup T_2)\cap[g(x,\theta_1)+g(x,\theta_2)>0]\cap[dP_{\theta_i}/dm(x)= g(x,\theta_i)h(x,B),\ i=1,2],$$

and $T_i=[dP_{\theta_i}/dm(x)>0]$, $i=1,2$.
Therefore it follows that

$$\int_{A\cap B} k(x)\,d(P_{\theta_1}+P_{\theta_2}) = \int_{C\cap B} g(x,\theta_1)/\{g(x,\theta_1)+g(x,\theta_2)\}$$

$$d(P_{\theta_1}+P_{\theta_2}) = \int_{C\cap B} \{dP_{\theta_1}/dm(x)\}/\{dP_{\theta_1}/dm(x)+dP_{\theta_2}/dm(x)\}$$

$$d(P_{\theta_1}+P_{\theta_2}) = \int_{C\cap B} dP_{\theta_1}/dm(x)\,dm = P_{\theta_1}(C\cap B)=P_{\theta_1}(A\cap B).$$

In particular we have

$$P_{\theta_1}(A)=P_{\theta_1}(A\cap(T_1\cup T_2))=\int_{A\cap(T_1\cup T_2)} k(x)\ d(P_{\theta_1}+P_{\theta_2})$$

$$=\int_A k(x)\,d(P_{\theta_1}+P_{\theta_2}).$$

By Lemma 2.1 (1) $\mathcal{B}$ is pairwise sufficient for $\mathcal{E}$.

<u>Remark 2.1</u>. Here we discuss the problem whether we can replace, in (2.3), the family of functions $\{h(.,A)\}$ with a single function. Suppose that there exists a function $h(.)$ on $\mathcal{X}$, which is not necessarily $\mathcal{A}$-measurable, such that

$$dP_\theta/dm(x)=g(x,\theta)h(x) \text{ a.e. } m \quad \text{for all } \theta\in\Theta, \tag{2.4}$$

where $g(.,\theta)$ is a non-negative $\mathcal{B}$-measurable function. Of course (2.4) should be interpreted that there exists an m-null set N_θ such that

$$[dP_\theta/dm(x) \neq g(x,\theta)h(x)] \subset N_\theta.$$

Define $T(\theta)=[dP_\theta/dm(x)>0]$, and $B(\theta)=[g(x,\theta)>0]$. Then for any $x\in(\mathcal{X}-N_\theta)\cap B(\theta)$, $h(x)=(dP_\theta/dm(x))/g(x,\theta)$ and $P_\theta(\mathcal{X}-(\mathcal{X}-N_\theta)\cap B(\theta))=0$ since $P_\theta(\mathcal{X}-B(\theta))=0$. Hence h is $\cap\{\bar{\mathcal{A}}(P_\theta);\theta\in\Theta\}$–measurable. Here $\bar{\mathcal{A}}(P_\theta)$ is the completed σ–algebra of $\mathcal{A}$ w.r.t. P_θ. If m is equivalent to $\mathcal{P}$ then this σ–field is equal to the σ–field

$$\{A\subset\mathcal{X}; \text{for any } F\in\mathcal{A}_\sigma(m) \text{ there exist } B \text{ and } N \text{ in } \mathcal{A} \text{ such that } (A\cap F)\Delta B\subset N, m(N)=0\}.$$

We denote this σ–field by $\tilde{\mathcal{A}}(m)$. Then the point is the following; even if $\mathcal{B}$ is PSS for $\mathcal{E}$ (2.4) does not necessarily hold.

We show this by applying an example of Fremlin(1978). Fremlin gave an example of a localizable measure space $(\mathcal{Z},\mathcal{C},\nu)$ for which $\mathcal{C}=\tilde{\mathcal{C}}(\nu)$, and is not strictly localizable. Fremlin (1978, Proposition 14) also proved that every measure space $(\mathcal{X},\mathcal{A},m)$ which satisfies $\tilde{\mathcal{A}}(m)=\mathcal{A}$ is strictly localizable if and only if its "complete, locally determined" product with every probability space is localizable. Hence there exists a probability space $(\mathcal{Y},\mathcal{B},\mu)$ such that the product $(\mathcal{Z}\times\mathcal{Y},\mathcal{C}\times\mathcal{B},\nu\times\mu)$ is not localizable even if we extend $\nu\times\mu$ to $\overline{\mathcal{C}\times\mathcal{B}}\,(\nu\times\mu)$, which we denote by $\overline{\mu\times\nu}$. Now let us define $\mathcal{X}=\mathcal{Z}\times\mathcal{Y}$, $\mathcal{A}=\overline{\mathcal{C}\times\mathcal{B}}\,(\nu\times\mu)$, $m'=\overline{\nu\times\mu}$, and $\mathcal{B}=\mathcal{C}\times\mathcal{Y}$. Then we have (1) $(\mathcal{X},\mathcal{A},m')$ is not localizable. (2) $(\mathcal{X},\mathcal{B},m'/\mathcal{B})$ is localizable, and (3) any maximal decomposition of $(\mathcal{X},\mathcal{B},m'/\mathcal{B})$ is also a maximal decomposition of $(\mathcal{X},\mathcal{A},m')$.

Since $(\mathcal{X},\mathcal{A},m')$ is not localizable, there exists a maximal decomposition $\{E_\gamma;\gamma\in\Gamma\}$ of $(\mathcal{X},\mathcal{A},m')$ and a set $\Gamma_1(\subset\Gamma)$ such that $\{E_\gamma;\gamma\in\Gamma_1\}$ does not have an essential supremum w.r.t. m (see Definition 1.4). Let us define a measure m on $(\mathcal{X},\mathcal{A})$ by

$$m(A)=m'(A)+\sum\{m'(A\cap E_\gamma);\gamma\in\Gamma_1\}.$$

Then m is equivalent to m', and there does not exist an $\mathcal{A}$-measurable function f, called a local density, such that

$$m'(A)=\int_A f(x)\ dm\ ,\ \text{whenever}\ m(A)<\infty\ .$$

If there exists such an f then $[f(x)<1]$ is an essential supremum of $\{E_\gamma;\gamma\in\Gamma_1\}$ w.r.t. m'.

Let us define $\mathcal{F}=\{F\in\mathcal{B};0<m'(F)<\infty\}$ and $\mathcal{P}=\{P_F;F\in\mathcal{F}\}$, where

$$P_F(A)=m'(A\cap F)/m'(F).$$

Then it can be proved that $\mathcal{P}\sim m$ because of (3) above, and $\mathcal{B}$ is PSS for $(\mathcal{X},\mathcal{A},\mathcal{P})$ since the subfield generated by $\mathcal{F}$ is the smallest PSS and $\mathcal{B}$ includes $\mathcal{F}$. Suppose that $dP_F/dm(x)$ can be factorized as

$$dP_F/dm(x)=g_F(x)h(x)\ \text{ a.e. } m,$$

where g_F is $\mathcal{B}$-measurable. Note that h is $\mathcal{A}$-measurable, because $\tilde{\mathcal{A}}(m)=\mathcal{A}$ and $\mathcal{P}\sim m$. Now from

$$dP_F/dm(x)=(1/m'(F))(dm'/dm)/F=g_F(x)h(x)\ \text{ a.e. } m,$$

we have $(dm'/dm)/F=m'(F)g_F(x)h(x)$ a.e. m. Hence $\{m'(F)g_F(x);F\in\mathcal{F}\}$ is a $\mathcal{B}$-measurable cross section w.r.t. m. Therefore, by Theorem 1.3, there exists a $\mathcal{B}$-measurable function g such that

$$I_F(x)g(x)=g_F(x)m'(F)\ \text{ a.e. } m,$$

because m is localizable on $(\mathcal{X},\mathcal{B})$. This shows

$$m'(F)=\int_F g(x)h(x)\ dm.$$

This can be extended to all $A\in\mathcal{A}(m)$. Hence gh is a local density of m' w.r.t. m. But this is a contradiction.

Conversely, however, if the family $\{h(x,A);A\in\mathcal{A}\}$ can be replaced by a single, not necessarily $\mathcal{A}$-measurable, function h on $\mathcal{X}$, then $\mathcal{B}$ is PSS for $\mathcal{E}$. Namely we have

<u>Theorem 2.4</u>. Suppose that

$$dP_\theta/dm(x)=g(x,\theta)h(x)\ \text{ a.e. } m, \tag{2.5}$$

for all θ , where $g(.,\theta)$ is a non-negative $\mathcal{B}$-measurable function and h is a non-negative function. Then $\mathcal{B}$ is PSS.

Proof. That $\mathcal{B}$ is pairwise sufficient is proved similarly to the proof of Theorem 2.3, so we omit the detail. We show $B_\theta=[g(x,\theta)>0]$ is a support of P_θ. Clearly $P_\theta(B_\theta)=1$. Take any $A\in\mathcal{A}$ such that $A\subset B_\theta$ and $P_\theta(A)=0$. Since h is $\cap\{\bar{\mathcal{A}}(P_\theta);\theta\in\Theta\}$-measurable as noted above, for any $Q\in\mathcal{P}$, there exist sets C_Q, $M_Q\in\mathcal{A}$ such that $[h(x)>0]\Delta C_Q\subset M_\theta$ and $Q(M_Q)=0$. We can take such a C_Q so as to satisfy $C_Q\supset[h(x)>0]$. Then

$$A\cap[dP_\theta/dm(x)>0]\cap(\mathcal{X}-N_\theta)=A\cap[h(x)>0]\cap(\mathcal{X}-N_\theta),$$

$P_\theta(A)=0$ imply that $m(A\cap[h(x)>0]\cap(\mathcal{X}-N_\theta))=0$, where $N_\theta\equiv N_{P_\theta}$ is an m-null set which includes $[dP_\theta/dm(x)\neq g(x,\theta)h(x)]$. So we have

$$Q(A\cap C_Q)=Q(A\cap C_Q\cap(\mathcal{X}-N_\theta))=Q(A\cap[h(x)>0]\cap(\mathcal{X}-N_\theta))=0.$$

On the other hand

$$\mathcal{X}-C_Q\subset[h(x)=0]\subset N_Q\cup[dQ/dm(x)=0]\in N(Q),$$

from which it follows that $Q(\mathcal{X}-C_Q)=0$. Thus we proved that $Q(A)=0$ and hence B_θ is a support of P_θ belonging to $\mathcal{B}$. This completes the proof.

2.3. Neyman factorization and pivotal measure in the case of weak domination

In this section we explain how the Neyman factorization theorem and pivotal measure discussed in the previous sections 2.1 and 2.2 are specialized in the case of weak domination. Experiment $\mathcal{E}=(\mathcal{X},\mathcal{A},\mathcal{P}=\{P_\theta;\theta\in\Theta\})$ considered in this section is assumed to be weakly dominated, and we suppose that m is a localizable majorizing measure for $\mathcal{E}$.

Theorem 2.5. A subfield $\mathcal{B}$ is PSS if and only if

$$dP_\theta/dm(x)=g(x,\theta)\ h(x)\ \text{a.e.}\ m\ \text{for all}\ \theta\in\Theta,$$

where $g(.,\theta)$ is a non-negative $\mathcal{B}$-measurable function and h is a non-negative $\mathcal{A}$-measurable function.

Proof. "If" part follows from Theorem 2.4.

Conversely suppose that $\mathcal{B}$ is PSS. Assume first that m is an equivalent localizable majorizing measure. Then $\{h(.,A);A\in\mathcal{A}_\sigma(m)\}$ is a cross section w.r.t. m by the proof of Theorem 2.3, and so it is determined by an $\mathcal{A}$-measurable function h because m has the finite subset property (Theorem 1.3). Then by Theorem 2.3 we have $dP_\theta/dm(x)=g(x,\theta)h(x)$ a.e.m on each $A\in\mathcal{A}_\sigma(m)$ and for all $\theta\in\Theta$. Hence we have $m(N_\theta\cap A)=0$ for all $A\in\mathcal{A}_\sigma(m)$, where $N_\theta=[dP_\theta/dm(x)\neq g(x,\theta)h(x)]$. Since m has the finite subset property this implies $m(N_\theta)=0$.

Next let m be any localizable majorizing measure. Let T be the essential supremum of $\{[dP_\theta/dm(x)>0];\theta\in\Theta\}$ w.r.t. m. Then m/T is a localizable measure on $(T,T\cap\mathcal{A})$ and is an equivalent majorizing measure of the experiment $(T,T\cap\mathcal{A},\mathcal{P}/T)$, where $\mathcal{P}/T=\{P_\theta/T;\theta\in\Theta\}$. It is easy to prove that $T\cap\mathcal{B}$ is PSS for this experiment. By the result above we have

$$d(P_\theta/T)/d(m/T)(x)=g'(x,\theta)h'(x) \text{ a.e. } m/T \text{ for all } \theta\in\Theta.$$

Define h by

$$h(x)=h'(x) \text{ if } x\in T, \text{ and}$$
$$=0 \quad \text{outside } T.$$

Take a $\mathcal{B}$-measurable function $g(.,\theta)$ extending $g'(.,\theta)$ to the whole space (Rao and Rao, 1981, p.9). Then we have

$$dP_\theta/dm(x)=g(x,\theta)h(x) \text{ a.e. } m.$$

This completes the proof.

Remark 2.2. Here let us see why such an extension as in Definition 1.2 of the notion of support is necessary. In the above proof we gave two cases of the proof separately depending on whether m is equivalent. If m is equivalent to $\mathcal{P}$ and $\mathcal{B}$ is PSS then each P_θ has a density $p(x,\theta)$ w.r.t. m such that $[p(x,\theta)>0]\in\mathcal{B}$. However if m is not equivalent to $\mathcal{P}$ then for some θ there may not exist any version of the density such that $[dP_\theta/dm(x)>0]\in\mathcal{B}$. For an example let $\mathcal{X}$ be the real line and $\mathcal{A}$ be the power set. Let $\mathcal{P}=\{P_\theta;\theta\in\mathcal{X},\theta\neq 0\}$ be the family of unit probability distribution P_θ on $\theta\in R^1,\theta\neq 0$. Then the counting measure m is localizable and is not equivalent to $\mathcal{P}$. Let $\mathcal{B}=\{A\in\mathcal{A};0,1\in A \text{ or } 0,1\bar{\in}A\}$. Then $S(\theta)=\{0,1\}$ if $\theta=1$, and $=\{\theta\}$ if $\theta\neq 0,1$, is a support of P_θ belonging to $\mathcal{B}$. It is easy to prove that $\mathcal{B}$ is PSS, and actually it is sufficient. Of course $[dP_1/dm(x)>0]\bar{\in}\mathcal{B}$.

Remark 2.3. There is a subfield for which the Neyman factorization holds in the sense of Theorem 2.10, but which is not sufficient. For an example let $\mathcal{X}$ be the real line, $\mathcal{A}$ be the power set, and consider the totality of unit distributions. Then the Borel field is PSS , and hence has the Neyman factorization. But it is not sufficient.

The latter half of this section is devoted to the study of pivotal measure in the case of weak domination. We will find a formula about pivotal measure which shows the equivalency, that is "equally informative", between the original experiment and the marginal experiment corresponding to the smallest PSS.

Lemma 2.2. If $\mathcal{B}$ is sufficient then it is PSS.

Proof. Let m be an equivalent majorizing measure for $\mathcal{E}$ and let f be the indicator of $T(\theta)=[dP_\theta/dm(x)>0]$. Then there exists a $\mathcal{B}$-measurable function g such that $0\leq g(x)\leq 1$ and

$$\int f(x)\,dQ=\int g(x)\,dQ \qquad \text{for all } Q\in\mathcal{P}.$$

Then $B(\theta)=[g(x)=1]\in\mathcal{B}$ satisfies that $T(\theta)\Delta\mathcal{B}(\theta)\in N(\mathcal{P})=N(m)$. Hence the function $f_\theta(x)$ defined by

$$\begin{aligned} f_\theta(x) &= dP_\theta/dm(x) && \text{if } x\in T(\theta)\cap B(\theta), \\ &= 1 && \text{if } x\in B(\theta)-T(\theta), \\ &= 0 && \text{otherwise,} \end{aligned}$$

satisfies that $f_\theta(x)=dP_\theta/dm(x)$ a.e. m and $[f_\theta(x)>0]=B(\theta)\in\mathcal{B}$.

Remark 2.4. Note that Lemma 2.2 holds true under the assumption that $\mathcal{E}$ is majorized and that $\mathcal{B}$ is test-sufficient. The proof and the definition of test sufficiency will be given in Chapter 4.

Next lemma shows how to obtain a sufficient σ-field from a pairwise sufficient subfield. For a subfield $\mathcal{B}$ its weak completion $\hat{\mathcal{B}}(\mathcal{A})$ is defined by

$$\{A\in\mathcal{A};\text{for any }\theta\in\Theta\text{ there exists }B_\theta\in\mathcal{B}\text{ such that }P_\theta(A\Delta B_\theta)=0\}.$$

Lemma 2.3. If $\mathcal{B}$ is pairwise sufficient then $\hat{\mathcal{B}}(\mathcal{A})$ is sufficient.

Proof. Since $\mathcal{B}$ is pairwise sufficient, by Lemma 2.1, for any $A\in\mathcal{A}$ the family of conditional probabilities $\{E(I_A/\mathcal{B},P_\theta);\theta\in\Theta\}$ is an essentially bounded countably coherent family. By Theorem 1.5, there exists an $\mathcal{A}$-measurable function f, such that

$$f(x)=E(I_A/\mathcal{B},P_\theta)(x)\text{ a.e. }P_\theta\text{ for all }\theta\in\Theta.$$

Then f is $\hat{\mathcal{B}}(\mathcal{A})$-measurable and each $E(I_A/\mathcal{B},P_\theta)$ is the conditional probability of A given $\hat{\mathcal{B}}(\mathcal{A})$ w.r.t. P_θ. Hence $\hat{\mathcal{B}}(\mathcal{A})$ is sufficient.

Next theorem shows that pivotal measure is characterized as an equivalent majorizing measure for which, for any sufficient subfield $\mathcal{B}$, each P_θ in $\mathcal{P}$ has a $\mathcal{B}$-measurable density.

Theorem 2.6. An equivalent majorizing measure n is pivotal for $\mathcal{E}$ if and only if for any sufficient subfield $\mathcal{B}$ and $\theta\in\Theta$ there exists a $\mathcal{B}$-measurable version of the density of P_θ w.r.t. n.

Proof. "Only if" part follows from Lemma 2.2. We show "if" part. Take any PSS subfield $\mathcal{B}$. Since n is equivalent to $\mathcal{P}$, as was shown in section 1.1, each P_θ has a density $q(.,\theta)$ w.r.t. n such that $[q(x,\theta)>0]\in\mathcal{B}$. On the other hand there exists a $\hat{\mathcal{B}}(\mathcal{A})$-measurable version $s(.,\theta)$ of the density of P_θ w.r.t. n by Lemma 2.3 and the assumption. Hence there exists a $\mathcal{B}$-measurable function $r(.,\theta)$ such that $r(x,\theta)=s(x,\theta)$ a.e. P_θ. Define a $\mathcal{B}$-measurable function $p(.,\theta)$ by

$$p(x,\theta)=s(x,\theta)\qquad\text{if }q(x,\theta)>0,\text{ and}$$

$= 0$ otherwise.

Then it is easy to check that $p(x,\theta)=q(x,\theta)$ a.e. n, which implies that $p(.,\theta)$ is a $\mathcal{B}$-measurable density of P_θ w.r.t. n.

Conversely take any subfield $\mathcal{B}$ such that each P_θ has a $\mathcal{B}$-measurable density w.r.t. n. Then by Neyman factorization theorem, Theorem 2.5, $\mathcal{B}$ is PSS. This completes the proof.

Lemma 2.4. If $\mathcal{B}$ is sufficient then $\hat{\mathcal{B}}(\mathcal{A})=\bar{\mathcal{B}}(\mathcal{A})$.

Proof. $\hat{\mathcal{B}}(\mathcal{A})\supset\bar{\mathcal{B}}(\mathcal{A})$ is clear. Take any A and B from $\hat{\mathcal{B}}(\mathcal{A})$. Then for any $\theta\in\Theta$ there exists $B_\theta\in\mathcal{B}$ such that $P_\theta(B\Delta B_\theta)=0$. Then we have, from

$$\int_B I_A dP_\theta = \int_{B_\theta} I_A dP_\theta = \int_{B_\theta} E(I_A/\mathcal{B})\,dP_\theta = \int_B E(I_A/\mathcal{B})\,dP_\theta,$$

$I_A(x)=E(I_A/\mathcal{B})(x)$ a.e. P_θ for all $\theta\in\Theta$. This shows that $A\in\bar{\mathcal{B}}(\mathcal{A})$.

If $\mathcal{E}$ is dominated then $\mathcal{E}(\mathcal{B})$ is also dominated for any subfield $\mathcal{B}$. But such a preservation of the structure is not true for weakly dominated experiments. As an example define $\mathcal{X}=$ the real line, $\mathcal{A}$ = the power set, and $\mathcal{P}=$ the totality of point probability measures. This experiment is weakly dominated, but the Borel field $\mathcal{B}$ is not sufficient, and $(\mathcal{X},\mathcal{B},\mathcal{P}/\mathcal{B})$ is not weakly dominated.

Next lemma gives a sufficient condition on $\mathcal{B}$ for which $\mathcal{E}(\mathcal{B})$ becomes weakly dominated.

Lemma 2.5. If $\mathcal{B}$ is sufficient then $\mathcal{E}(\mathcal{B})$ is also weakly dominated.

Proof. Let $\{f(x,\theta);\theta\in\Theta\}$ be any essentially bounded countably coherent family of $\mathcal{B}$-measurable functions. Then there exists an $\mathcal{A}$-measurable function f such that $f(x)=f(x,\theta)$ a.e. P_θ for all $\theta\in\Theta$, because $\mathcal{E}$ is weakly dominated (Theorem 1.5). Hence f is $\hat{\mathcal{B}}(\mathcal{A})$-measurable. Since $\mathcal{B}$ is sufficient $\hat{\mathcal{B}}(\mathcal{A}) = \bar{\mathcal{B}}(\mathcal{A})$ by Lemma 2.4. Hence there exists a $\mathcal{B}$-measurable function h such that

$$h(x)=f(x)=f(x,\theta) \text{ a.e. } P_\theta \text{ for all } \theta\in\Theta.$$

Again by Theorem 1.5 $\mathcal{E}(\mathcal{B})$ is weakly dominated.

Next theorem shows that there exists the smallest sufficient subfield in the case of weak domination.

Theorem 2.7. Let $\mathcal{B}_0$ be the smallest PSS. Then $\hat{\mathcal{B}}_0(\mathcal{A})$ is the smallest sufficient subfield.

Proof. By Lemma 2.3, $\hat{\mathcal{B}}_0(\mathcal{A})$ is sufficient. Take any sufficient subfield $\mathcal{C}$. Then we have $\mathcal{B}_0 \subset \mathcal{C}[\mathcal{P}]$ because $\mathcal{C}$ is PSS by Lemma 2.2. Therefore it holds that $\bar{\mathcal{C}}(\mathcal{A}) = \hat{\mathcal{C}}(\mathcal{A}) \supset \hat{\mathcal{B}}_0(\mathcal{A})$ by Lemma 2.4, which implies $\mathcal{C} \supset \hat{\mathcal{B}}_0(\mathcal{A})\ [\mathcal{P}]$. Hence $\hat{\mathcal{B}}_0(\mathcal{A})$ is the smallest sufficient subfield.

Now, as was shown in the proof of Theorem 2.2, for any maximal decomposition $\mathcal{F}=\{F\}$ of the majorized experiment $\mathcal{E}(\mathcal{B}_0) = (\mathcal{X}, \mathcal{B}_0, \mathcal{P}/\mathcal{B}_0)$ the measure $n(A) = \sum\{P_F(A \cap F); F \in \mathcal{F}\}$, $A \in \mathcal{A}$, is a pivotal measure for $\mathcal{E}$. Since $\mathcal{E}$ is weakly dominated this n is localizable by Lemma 1.3 and Definition 1.4. Moreover we have

$$
\begin{aligned}
n(A) &= \sum\{P_F(A \cap F); F \in \mathcal{F}\} \\
&= \sum\{\textstyle\int_F E(I_A/\hat{\mathcal{B}}_0(\mathcal{A}))\ dP_F; F \in \mathcal{F}\} && (\hat{\mathcal{B}}_0(\mathcal{A}) \text{ is sufficient}) \\
&= \sum\{\textstyle\int_{\mathcal{X}} E(I_A/\hat{\mathcal{B}}_0(\mathcal{A}))\, dQ_F; F \in \mathcal{F}\} && (dQ_F \equiv I_F dP_F) \\
&= \textstyle\int_{\mathcal{X}} E(I_A/\mathcal{B}_0(\mathcal{A}))\, d(\sum\{Q_F; F \in \mathcal{F}\}).
\end{aligned}
$$

Let us define a measure n' on $(\mathcal{X}, \hat{\mathcal{B}}_0(\mathcal{A}))$ by

$$n'(B) = \sum\{Q_F(B); F \in \mathcal{F}\}.$$

Then we have $\mathcal{P}/\hat{\mathcal{B}}_0(\mathcal{A}) \sim n'$ and n' has the finite subset property. By Lemma 2.5 $\mathcal{E}(\hat{\mathcal{B}}_0(\mathcal{A}))$ is weakly dominated, so n' is localizable. Let us take any equivalent majorizing measure m' for $\mathcal{E}(\hat{\mathcal{B}}_0(\mathcal{A}))$. Then any σ-finite set w.r.t. n' is also σ-finite w.r.t. m', and vice versa by Lemma 1.3. This implies that n' is an equivalent majorizing measure for $\mathcal{E}(\hat{\mathcal{B}}_0(\mathcal{A}))$. Let us write down again that

$$n(A)=\int_{\mathcal{X}} E(I_A/\hat{\mathcal{B}}_0(\mathcal{A}))\,dn'$$

$$=\int_{\mathcal{X}} E(I_A/\hat{\mathcal{B}}_0(\mathcal{A}))\,dn/\hat{\mathcal{B}}_0(\mathcal{A}).$$

Note that this is the same representation as that of randomization obtaining $\mathcal{E}$ from $\mathcal{E}(\hat{\mathcal{B}}_0(\mathcal{A}))$. This is indeed characteristic of the pivotal measure. Namely for any equivalent localizable majorizing measure n' for $\mathcal{E}(\hat{\mathcal{B}}_0(\mathcal{A}))$ the measure n defined by

$$n(A)=\int_{\mathcal{X}} E(I_A/\hat{\mathcal{B}}_0(\mathcal{A}))\ dn',\ A\in\mathcal{A},$$

is a pivotal measure for $\mathcal{E}$. This can be shown by Theorem 2.6. But we omit the detail.

2.4. Dominated case

In this section we go back to the Halmos-Savage-Bahadur theory by applying results of previous sections.

Let $\mathcal{E}=(\mathcal{X},\mathcal{A},\mathcal{P}=\{P_\theta;\theta\in\Theta\})$ be dominated by a σ-finite measure m. Let us denote $S(\theta)=[dP_\theta/dm(x)>0]$ for each $\theta\in\Theta$. Let $\mathcal{F}=\{F\}$ be the maximal decomposition of $\mathcal{E}$ defined in section 1.1. Then

$$n(A)=\sum\{P_F(A\cap F);F\in\mathcal{F}\},\ A\in\mathcal{F},$$

is an equivalent dominating measure of $\mathcal{E}$ by Theorem 1.1. Hence it has the finite subset property by Lemma 1.2. Then by Lemma 1.3 n itself is a σ-finite measure. By Proposition 1.1 $\{F\in\mathcal{F};\ n(F)>0\}=\mathcal{F}$ is at most countable. Let us denote it by $\mathcal{F}=\{F_1,F_2,\ldots\}$. Then for any $i\geq 1$ there exists P_{θ_i} such that $P_{\theta_i}(F_i)>0$ and $F_i\subset S(\theta_i)$. Then for $\mathcal{P}_0\equiv\{P_{\theta_1},P_{\theta_2},\ldots\}$ we have $N(\mathcal{P}_0)=N(\mathcal{P})$. Hence,

Theorem 2.8. There exists a countable subfamily $\mathcal{P}_0$ of $\mathcal{P}$ such that $N(\mathcal{P}_0)=N(\mathcal{P})$.

Corollary 2.1. There exists an equivalent finite dominating measure for $\mathcal{E}$.

For example, $\sum 2^{-i}P_{\theta_i}$.

Theorem 2.9. If $\mathcal{B}$ is pairwise sufficient then it is sufficient.

Proof. By Lemma 2.3 $\hat{\mathcal{B}}(\mathcal{A})$ is sufficient. By Lemma 2.1 we have $\hat{\mathcal{B}}(\mathcal{A})=\{A\in\mathcal{A};$ for any countable subfamily $\mathcal{P}_0$ of $\mathcal{P}$ there exists B in $\mathcal{B}$ such that $A\Delta B\in N(\mathcal{P}_0)\}$. By the definition of sufficiency, for any $A\in\mathcal{A}$, there exists a $\hat{\mathcal{B}}(\mathcal{A})$-measurable function g such that

$$P_\theta(A\cap C)=\int_C h(x)\,dP_\theta \quad \text{for all } C\in\hat{\mathcal{B}}(\mathcal{A}) \text{ and } \theta\in\Theta.$$

By Theorem 2.8 there exists a countable subfamily $\mathcal{P}_0\subset\mathcal{P}$ such that $N(\mathcal{P}_0)=N(\mathcal{P})$. Hence for this g there exists a $\mathcal{B}$-measurable function h such that $g(x)=h(x)$ a.e. P_θ for all $\theta\in\Theta$. Then we have

$$P_\theta(A\cap B)=\int_B h(x)\;dP_\theta \text{ for all } B\in\mathcal{B} \text{ and } \theta\in\Theta,$$

which implies $\mathcal{B}$ is sufficient.

Corollary 2.2. $\mathcal{B}$ is sufficient if and only if it is PSS.

Let us define a finite measure n by $n(A)=\sum_{i=1}^{\infty} c_i P_{\theta_i}(A)$, where $A\in\mathcal{A}$, $c_i>0$, $\sum_{i=1}^{\infty} c_i<\infty$. Here $\mathcal{P}_0=\{P_{\theta_1},P_{\theta_2},\ldots\}$ is a countable subfamily of $\mathcal{P}$ such that $N(\mathcal{P}_0)=N(\mathcal{P})$. Let $\mathcal{B}_0$ be the smallest sufficient subfield whose existence is guaranteed in the more general case of weak domination (Theorem 2.7). Then the measure

$$n'(A)=\int_{\mathcal{X}} E(I_A/\mathcal{B}_0)\,dn/\mathcal{B}_0\;,\quad A\in\mathcal{A},$$

is a pivotal measure from the discussion in the final paragraph of section 2.3. But we have

$$n'(A)=\int_{\mathcal{X}} E(I_A/\mathcal{B}_0)\,dn/\mathcal{B}_0$$

$$=\sum_{i=1}^{\infty} c_i \int_{\mathcal{X}} E(I_A/\mathcal{B}_0)\, dP_{\theta_i}/\mathcal{B}_0$$

$$=\sum_{i=1}^{\infty} c_i P_{\theta_i}(A)=n(A).$$

Therefore n itself is a pivotal measure for $\mathcal{E}$.

As for Neyman factorization theorem in the case of domination we have

Theorem 2.10. $\mathcal{B}$ is sufficient if and only if

$$dP_\theta/dm(x)=g(x,\theta)h(x) \text{ a.e. } m \text{ for all } \theta\in\Theta,$$

where $g(.,\theta)$ is a non-negative $\mathcal{B}$-measurable function and h is a non-negative $\mathcal{A}$-measurable function.

Proof. Suppose that $\mathcal{B}$ is sufficient. Since n is a pivotal measure and by Theorem 2.6 each P_θ has a $\mathcal{B}$-measurable version $g(.,\theta)$ of the density w.r.t. n. Then we have

$$dP_\theta/dm(x)=g(x,\theta)\,dn/dm(x)=g(x,\theta)h(x) \text{ a.e. } m.$$

The converse part of the Theorem follows from Theorem 2.5 and Corollary 2.2.

Corollary 2.3. $\mathcal{B}$ is sufficient if and only if there exists an equivalent finite dominating measure w.r.t. which each P_θ in $\mathcal{P}$ has a $\mathcal{B}$-measurable density.

2.5. Bibliographical notes

The notions of PSS and pivotal measure in undominated cases were introduced by Ghosh, Morimoto and Yamada(1981). This paper also contains results on the smallest PSS and many examples. The nomenclature "pivotal" was used by Bahadur(1957) for the first time. Pivotal measure in majorized experiment was constructed by Ramamoorthi and Yamada(1982). Remark 2.1 is also due to this paper. Neyman factorization theorem was originally proved by Fisher (1922), and later by Neyman (1935) rigorously. So sometimes this theorem is called Fisher-Neyman factorization theorem. The fact that Neyman factorization does not necessarily imply sufficiency was discovered by Morimoto (1972). Yamada(1982) is a survey paper to which Section 2.4 was referred.

Lemma 2.3 is very important. It is applied many times. Naturally it was used to show equivalency of pairwise sufficiency to sufficiency in dominated cases (Theorem 2.9). For a similar proof of Theorem 2.9 we refer to Speed(1976). It is interesting, related to Lemma 2.3, to know under what conditions pairwise sufficiency implies sufficiency. Kusama and Yamada(1972), Speed(1976) showed the following: Let $\mathcal{E}=(\mathcal{X},\mathcal{A},\mathcal{P})$ be any experiment and $\mathcal{B}$ be pairwise sufficient. Then $\mathcal{B}$ is sufficient if and only if $\mathcal{E}(\mathcal{B})$ is weakly dominetd. As for conditions for a PSS subfield to be sufficient we refer to Fujii and Morimoto(1985).

Chapter 3. Structure of pairwise sufficient subfield and PSS

Lehmann and Scheffe (1950) gave a concrete construction of the smallest sufficient statistic, which is generated by the equivalent likelihood partition of the sample space, under the assumption of separabilty of the family of the density functions. On the other hand Bahadur (1954) constructed the smallest sufficient σ-field for any dominated experiment as the one generated by the densities w.r.t. the dominating measure $n=\sum_{i=1}^{\infty} c_i P_{\theta_i}$ introduced in Section 2.4.

In this chapter we will give a geometric representation of pairwise sufficient subfield and PSS. In Section 3 we also study the order preserving of sufficiency related to the Burkholder problem. We first introduce the Lehmann and Scheffé type characterization of the smallest sufficient statistic in the case of discrete experiment. Then from this result we induce an essential property of the smallest PSS for discrete experiments, and then assert that this property indeed characterizes the smallest PSS subfield for general majorized experiments. Basu and Ghosh(1967), Morimoto(1972), Ghosh, Morimoto and Yamada(1981), and, Fujii and Morimoto(1986) are main references for this chapter.

3.1. Discrete experiment case

Let $\mathcal{E}=(\mathcal{X},\mathcal{A},\mathcal{P}=\{P_\theta;\theta\in\Theta\})$ be a discrete experiment. That is, $\mathcal{X}$ is an uncountable set, $\mathcal{A}$ is the power set of $\mathcal{X}$ and each P_θ is discrete such that $N(\mathcal{P})=\{\emptyset\}$. Discrete experiment is weakly dominated by the counting measure on $(\mathcal{X},\mathcal{A})$. Let us denote by $p(x,\theta)$ the density of P_θ w.r.t. the counting measure. Although the aim of this section is to give characterization of pairwise sufficiency and PSS, we would like to represent some related basic results here, including a Neyman factorization theorem, because of its interesting speciality.

A *statistic*, by its definition, is a partition $\mathcal{T}=\{T\}$ of mutually disjoint, non-empty subsets of $\mathcal{X}$ which collectively covers $\mathcal{X}$. Then each $T\in\mathcal{T}$ is called a *part*. For two statistics $\mathcal{T}$ and $\mathcal{U}$ we write $\mathcal{T}>\mathcal{U}$ and read that $\mathcal{T}$ is *finer* than $\mathcal{U}$ if any part of $\mathcal{U}$ is a union of some parts of $\mathcal{T}$. For a statistic $\mathcal{T}=\{T\}$ let $\mathcal{B}(\mathcal{T})$ be the subfield consisting of sets which are unions of parts of $\mathcal{T}$. $\mathcal{B}(\mathcal{T})$ is called the subfield induced by $\mathcal{T}$. If $\mathcal{B}(\mathcal{T})$ is (pairwise) sufficient then we say, by the definition, that $\mathcal{T}$ is (pairwise) sufficient.

To prove the following Neyman factorization theorem for a statistic it is easier to give a direct proof than to apply the general results of the previous chapter. On the contrary, at some stage of the proof the general results in Chapter 2 are not enough to apply. But in Chapter 4 we will give a general result (Theorem 4.5) to show applications to a few important results on sufficiency in the discrete case including the Neyman factorization theorem. A direct proof of the following theorem works well because we consider discrete distributions, (x) belongs to $\mathcal{A}$ for each $x\in\mathcal{X}$, and $\mathcal{B}(\mathcal{T})$ is closed under the formation of union of any number of sets in it.

Theorem 3.1. A statistic $\mathcal{T}=\{T\}$ is sufficient if and only if it holds that

$$p(x,\theta)=p(T_x,\theta)h(x) \tag{3.1}$$

for all $x\in\mathcal{X}$ and $\theta\in\Theta$, where $x\in T_x\in\mathcal{T}$, $p(T_x,\theta)=P_\theta(T_x)$ and $h(x)$ is independent of θ.

Proof. Suppose that $\mathcal{T}$ is sufficient. Since $\mathcal{B}(\mathcal{T})$ is sufficient we have

$$p(x,\theta)=\sum\{E(I_{\{x\}}/\mathcal{B}(\mathcal{T}))(y)p(y,\theta);y\in T_x\}=p(T_x,\theta)h(x).$$

Since $E(I_{\{x\}}/\mathcal{B}(\mathcal{T}))(y)$, as a function of y, is $\mathcal{B}(\mathcal{T})$-measurable it is constant on each part of $\mathcal{T}$. The constant value on T_x is denoted by $h(x)$. Then we have (3.1).

Conversely let us assume the factorization (3.1). We note that $h(x)>0$ for all $x\in\mathcal{X}$ because $N(\mathcal{P})=\{\emptyset\}$ holds by the assumption. Take any $T\in\mathcal{T}$. For any $x\in T$ we have from (3.1) $p(x,\theta)=P_\theta(T)h(x)$ for all $\theta\in\Theta$. Since $T\neq\emptyset$ there exists at least one $\theta\in\Theta$ such that $P_\theta(T)>0$. Take any such a θ. Then we have $p(x,\theta)>0$ for all $x\in T$. Hence T itself is countable. Take any $A\in\mathcal{A}$ and define a function $f(A,x)$ by

$$f(A,x) = \sum\{h(y);\ y\in A\cap T_x\},$$

where T_x is the part of $\mathcal{T}$ containing x. Then $f(A,x)$ is constant on each part of $\mathcal{T}$. So $f(A,x)$, as a function of x, is $\mathcal{B}(\mathcal{T})$-measurable. The constant value of $f(A,x)$ on $T\in\mathcal{T}$ is denoted by $f(A,T)$. Hence $f(A,T)=\Sigma\{h(y);y\in A\cap T\}$. Then for any $B\in \mathcal{B}(\mathcal{T})$ and $\theta\in\Theta$ we have

$$\int_B f(A,x)\ dP_\theta=\Sigma\{f(A,x)p(x,\theta);\ x\in B\}$$

$$=\Sigma\{f(A,T)P_\theta(T);T\subset B,T\in\mathcal{T}\}$$

$$=\Sigma\{\{\Sigma\{h(y);\ y\in A\cap T\}\}P_\theta(T);T\subset B,T\in\mathcal{T}\}$$

$$=\Sigma\{\Sigma\{p(y,\theta);y\in A\cap T\};T\subset B,T\in\mathcal{T}\}$$

$$=\Sigma\{p(x,\theta);y\in A\cap B\}\ =\ P_\theta(A\cap B).$$

This shows that $f(A,x)$ is the conditional probability of A given $\mathcal{B}(\mathcal{T})$ which is common to all $P_\theta\in\mathcal{P}$. Hence $\mathcal{B}(\mathcal{T})$ is sufficient. Namely $\mathcal{T}$ is sufficient .

We define a notion of "induced by a statistic" of a subfield. This becomes important when we (adhere to sufficiency to) try to overcome the Burkholder pathology. A subfield $\mathcal{B}$ is inducible if there exists a statistic $\mathcal{T}$ such that $\mathcal{B}=\mathcal{B}(\mathcal{T})$. Then we say $\mathcal{B}$ is induced by $\mathcal{T}$. We note that for any statistic $\mathcal{T}$ it follows that $\mathcal{T}=\mathcal{T}(\mathcal{B}(\mathcal{T}))$. Here $\mathcal{T}(\mathcal{B})$ is the statistic induced by $\mathcal{B}$ which is defined by the following equivalence relation. For any x and y in $\mathcal{X}$ x is equivalent to y, denoted by $x\sim y$, if and only if each B in $\mathcal{B}$ contains both x and y or neither of them. So any statistic $\mathcal{T}$ is always induced by $\mathcal{B}(\mathcal{T})$.

Theorem 3.2. Any sufficient subfield $\mathcal{B}$ is induced by a sufficient statistic $\mathcal{T}$.

Proof. Let's define $\mathcal{T}=\mathcal{T}(\mathcal{B})$. Take any $T\in\mathcal{T}$ and put

$$f(x)=E(I_T/\mathcal{B})(x).$$

Since f is $\mathcal{B}$-measurable it is constant on each part of $\mathcal{T}$. The constant which f takes on T is denoted by c. Then $C\equiv[f(x)=c]$ satisfies $C\in\mathcal{B}$ and $T\subset C$. We will show $T\in\mathcal{B}$ by proving $T=C$. At first we note that the constant c is positive. This follows from the fact that there exists $\theta \in \Theta$ such that $P_\theta(T)>0$, and

$$P_\theta(T)=P_\theta(T\cap C)=\int_C f(x)\,dP_\theta(x)=c\cdot P_\theta(C).$$

To prove $T=C$ it is enough to show the following; for any $T_1 \in \mathcal{T}$ such that $T_1 \neq T$ f takes a constant on T_1 different from c. If f takes the same constant c on T_1 then we have $T_1 \subset C$. Let's take $B \in \mathcal{B}$ such that $T_1 \subset B$ and $T \cap B=\emptyset$. Then for any $\theta \in \Theta$ we have

$$0=P_\theta(T\cap B\cap C)=\int_{B\cap C} f(x)\,dP_\theta(x)=cP_\theta(B\cap C)\geq cP_\theta(T),$$

which implies $P_\theta(T_1)=0$ for all $\theta \in \Theta$. This is a contradiction.

Next we show any $B \in \mathcal{B}(\mathcal{T})$ belongs to $\mathcal{B}$. For this purpose let's consider the conditional probability g of B given $\mathcal{B}$, $g(x)=E(I_B/\mathcal{B})(x)$. Take any $T \in \mathcal{T}$ and let d be the constant which g takes on T. Since $T \in \mathcal{B}$ we have

$$P_\theta(T\cap B)=\int_T g(x)\,dP_\theta(x)=dP_\theta(T)>0$$

for some $\theta \in \Theta$. This shows d is 1 or 0 depending on $B \supset T$ or $B \cap T=\emptyset$ respectively. Since B is a union of some parts of $\mathcal{T}$ we have $g(x)=I_B(x)$, so $B \in \mathcal{B}$. Thus we proved $\mathcal{B}(\mathcal{T}(\mathcal{B}))\subset \mathcal{B}$.

Since any $B \in \mathcal{B}$ can be represented as $B=\cup\{T\in\mathcal{T}(\mathcal{B});T\subset B\}$, the converse inclusion $\mathcal{B}(\mathcal{T}(\mathcal{B}))\supset\mathcal{B}$ is obvious. Thus we have $\mathcal{B}=\mathcal{B}(\mathcal{T}(\mathcal{B}))$.By the definition $\mathcal{T}(\mathcal{B})$ is sufficient.

<u>Corollary 3.1</u>. Any sufficient subfield is closed under the formation of union of any number of sets in it.

This follows from Theorem 3.2 and suggests that the Burkholder phenomena may occur in the discrete case because there are subfields which contain sufficient subfields and are not closed with respect to taking union of any number of sets in them. In Section 3.3 we will study this Burkholder problem.

<u>Definition 3.1</u>. A subfield which is closed with respect to union of any number of sets in it is said to be *closed*.

<u>Proposition 3.1</u>. A subfield $\mathcal{B}$ is inducible if and only if it is closed.

Proof. If $\mathcal{B}$ is induced by a ststistic then it is clear that $\mathcal{B}$ is closed. To show the converse take any $x\in\mathcal{X}$ and define $T_x=\cap\{B\in\mathcal{B};x\in B\}$. Then $\{T_x;x\in\mathcal{X}\}$ defines a statistic $\mathcal{T}$ such that $T_x\in\mathcal{B}$ for all $x\in\mathcal{X}$. Since $\mathcal{B}$ is closed it follows that $\mathcal{B}(\mathcal{T})\subset\mathcal{B}$. Let's take $B\in\mathcal{B}$. Since $B=\cup\{T_x;x\in B\}$ we have $\mathcal{B}\subset\mathcal{B}(\mathcal{T})$. We proved $\mathcal{B}=\mathcal{B}(\mathcal{T})$, so $\mathcal{B}$ is inducible.

A sufficient statistic $\mathcal{T}$ is called the *smallest sufficient statistic* if $\mathcal{T}<\mathcal{U}$ for every sufficient statistic $\mathcal{U}$.

Theorem 3.3. A sufficient statistic $\mathcal{T}$ is the smallest sufficient statistic if and only if $\mathcal{B}(\mathcal{T})$ is the smallest sufficient subfield.

Proof. Suppose that $\mathcal{T}$ is the smallest sufficient statistic. Take any sufficient subfield $\mathcal{C}$. By Theorem 3.2, $\mathcal{C}=\mathcal{B}(\mathcal{T}(\mathcal{C}))$ and $\mathcal{T}(\mathcal{C})$ is a sufficient statistic. Then $\mathcal{T}(\mathcal{C})>\mathcal{T}$ implies $\mathcal{C}=\mathcal{B}(\mathcal{T}(\mathcal{C}))\supset\mathcal{B}(\mathcal{T})$. Hence $\mathcal{B}(\mathcal{T})$ is the smallest sufficient subfield.

Conversely let $\mathcal{B}(\mathcal{T})$ be the smallest sufficient subfield. Take any sufficient statistic $\mathcal{U}$. Then we have $\mathcal{B}(\mathcal{U})\supset\mathcal{B}(\mathcal{T})$. For any $T\in\mathcal{T}$, $T\in\mathcal{B}(\mathcal{U})$ implies T is a union of parts of $\mathcal{U}$, which shows $\mathcal{U}>\mathcal{T}$. Hence $\mathcal{T}$ is the smallest sufficient statistic.

Next theorem shows how to construct the smallest sufficient statistic, and it is the same to the Lehmann and Scheffé method. Note that in this case the assumption of the separability is not necessary, because we are taking the power set as the basic σ-field.

Theorem 3.4. There exists the smallest sufficient statistic $\mathcal{M}$. Each part of $\mathcal{M}$ is defined by the following equivalence relation; for any x and y from $\mathcal{X}$, $x\sim y$ if and only if,

(1) $p(x,\theta)>0$ implies $p(y,\theta)>0$ and vice versa for all θ, and

(2) $p(x,\theta)/p(y,\theta)$ is independent of θ for which $p(x,\theta)p(y,\theta)>0$ holds.

Proof. For any $x\in\mathcal{X}$ let us define P_x by $P_x=\{\theta\in\Theta;p(x,\theta)>0\}$. By the definition of discrete experiment P_x is not empty. We define an equivalence relation $x \sim y$ on $\mathcal{X}$ by the conditions that $\mathcal{P}_x = \mathcal{P}_y$ and $p(x,\theta)/p(y,\theta)$ is a constant in θ for all $\theta \in \mathcal{P}_x$. Take any $x \in \mathcal{X}$ and define $\pi_x = \{y; y \sim x\}$. Then $p(y,\theta)/p(x,\theta)$ is a constant in θ for all $y \in \pi_x$, and so $P_\theta(\pi_x)/p(x,\theta)$ is also a constant in θ. If we put

$$h(x) = p(x,\theta)/P_\theta(\pi_x)$$

it follows that

$$p(x,\theta) = P_\theta(\pi_x)h(x)$$

for all $x \in \mathcal{X}$ and $\theta \in \Theta$. By Theorem 3.1 $\mathfrak{M} = \{\pi_x; x \in \mathcal{X}\}$ is a sufficient statistic. Take any sufficient statistic $\mathcal{U}$. If x and y belong to the same part in $\mathcal{U}$ we have $x \sim y$ by Theorem 3.1 again. This implies that $\mathcal{U} > \mathfrak{M}$.

For any statistic $\mathcal{T}$ let us denote by $\sigma(\mathcal{T})$ the subfield which is generated by $\mathcal{T}$. Namely it is the totality of sets consisting of countable unions of parts of $\mathcal{T}$ and their complements.

Theorem 3.5. If $\mathcal{T}$ is a sufficient statistic then $\sigma(\mathcal{T})$ is pairwise sufficient.

Proof. By Theorem 3.4 each $M \in \mathfrak{M}$ is countable, since it is contained in $[p(x,\theta) > 0]$ for some $\theta \in \Theta$. Then $\mathcal{T} > \mathfrak{M}$ implies that M is a countable union of parts in $\mathcal{T}$, and this shows $\sigma(\mathcal{T}) \supset \sigma(\mathfrak{M})$. Hence it is enough to show that $\sigma(\mathfrak{M})$ is pairwise sufficient. Take any $\theta_1, \theta_2 \in \Theta$, and define

$$A = [p(x,\theta_1)p(x,\theta_2) > 0], B = [p(x,\theta_1) > 0,\ p(x,\theta_2) = 0].$$

Then

$$dP_{\theta_1}/dm(x) = \begin{cases} 1/(1 + p(x,\theta_1)/p(x,\theta_2)) & \text{if } x \in A, \\ 1 & \text{if } x \in B, \\ 0 & \text{otherwise,} \end{cases}$$

where $m = P_{\theta_1} + P_{\theta_2}$. Clearly A and B are countable unions of sets in $\mathfrak{M}$. Hence they belong to $\sigma(\mathfrak{M})$. For any $0 < \alpha < 1$ it holds that $[dP_{\theta_1}/dm(x) \leq \alpha] \in \sigma(\mathfrak{M})$. Hence $dP_{\theta_1}/dm(x)$ is $\sigma(\mathfrak{M})$-measurable.

Similarly $dP_{\theta_2}/dm(x)$ is $\sigma(\mathfrak{m})$-measurable. These imply that $\sigma(\mathfrak{m})$ is sufficient for $\{P_{\theta_1},P_{\theta_2}\}$ by the Neyman factorization theorem in a dominated case.

Next result shows that any pairwise sufficient subfield separates the smallest sufficient statistic. Take two families of sets $\mathcal{B}$ and $\mathcal{C}$ from $\mathcal{A}$. Then it is said that $\mathcal{B}$ *separates* $\mathcal{C}$ if, for any different C_1 and $C_2\in\mathcal{C}$, there exists $B\in\mathcal{B}$ such that both C_1-B and $C_2\cap B$ belong to $N(\mathcal{P})$.

Theorem 3.6. Suppose that $\mathcal{B}$ is pairwise sufficient. Then it holds $\mathcal{T}(\mathcal{B})>\mathcal{T}(\sigma(\mathfrak{m}))=\mathfrak{m}$. Then $\mathcal{B}$ separates $\mathfrak{m}$.

Proof. Take any two points x and y which are not $x\sim y$ w.r. t. $\mathfrak{m}$. Then by Theorem 3.4 there exist $\theta_1,\theta_2\in\Theta$ such that

$$p(x,\theta_2)/p(x,\theta_1)>p(y,\theta_2)/p(y,\theta_1).$$

Define α as the number

$$\Sigma\{p(z,\theta_1);p(z,\theta_2)/p(z,\theta_1)>p(y,\theta_2)/p(y,\theta_1)\}.$$

Then there exists a $\mathcal{B}$-measurable most powerful level α test of $p(x,\theta_1)$ against $p(x,\theta_2)$ by the assumption that $\mathcal{B}$ is sufficient for $\{P_{\theta_1},P_{\theta_2}\}$. From the fundamental lemma of Neyman and Pearson this test has value 1 at x, and 0 at y. Hence there exists $B\in\mathcal{B}$ such that $x\in B$ and $y\bar{\in}B$. This shows $\mathcal{T}(\mathcal{B})>\mathfrak{m}$.

As for the latter part let us define

$$M_i=\bigcup_{j=1}^{\infty}T_j^i,\quad T_j^i\in\mathcal{T}(\mathcal{B}),\ i=1,2.$$

Then there exists $B_{ij}\in\mathcal{B}$ such that $B_{ij}\supset T_i^1$ and $B_{ij}\cap T_j^2=\emptyset$ for each i and j. Then $B=\bigcap_{j=1}^{\infty}\bigcup_{i=1}^{\infty}B_{ij}\in\mathcal{B}$ satisfies $B\supset M_1$ and $B\cap M_2=\emptyset$.

Next theorem gives a sufficient condition for pairwise sufficiency. In any discrete experiment $[p(x,\theta)>0]$ is the unique support of P_θ.

Theorem 3.7. Suppose that $\mathcal{B}$ contains supports and $\mathcal{T}(\mathcal{B})>\mathfrak{m}$. Then we have $\mathcal{B}\supset\sigma(\mathfrak{m})$. Hence in particular $\mathcal{B}$ is pairwise sufficient.

<u>Proof</u>. By the assumption that $S(P_\theta)$ belongs to $\mathcal{B}$ for all $\theta\in\Theta$ each $T\in\mathcal{T}(\mathcal{B})$ is contained in some $S(P_\theta)$. Thus $S(P_\theta)$ can be written as $T\cup T_1\cup T_2\cup\ldots$, where $T_i\in\mathcal{T}(\mathcal{B})$, $i=1,2,\ldots$. By Theorem 3.6, for any $i\geq 1$, there exists a set $B_i\in\mathcal{B}$ such that $B_i\supset T$ and $B_i\cap T_i=\emptyset$. Then we have $T=(\bigcap_{i=1}^{\infty}B_i)\cap S(P_\theta)$, from which it follows $T\in\mathcal{B}$. Thus we proved $\mathcal{T}(\mathcal{B})\subset\mathcal{B}$. Take any $M\in\mathfrak{M}$. By Theorem 3.4, there exists $\theta\in\Theta$ such that $M\subset S(P_\theta)$ and hence M is a countable set. $\mathcal{T}(\mathcal{B})>\mathfrak{M}$ implies that M is a countable union of parts in $\mathcal{T}(\mathcal{B})$. So $M\in\mathcal{B}$ holds by $\mathcal{T}(\mathcal{B})\subset\mathcal{B}$. Therefore we have $\mathcal{B}\supset\mathfrak{M}$ and $\mathcal{B}\supset\sigma(\mathfrak{M})$.

To give the main result in this section we need still two more lemmas.

<u>Lemma 3.1</u>. A subfield $\mathcal{B}$ is pairwise sufficient if and only if it separates $\mathfrak{M}$.

<u>Proof</u>. "Only if" part. By Theorem 3.6 we have $\mathcal{T}(\mathcal{B})>\mathfrak{M}$. Hence any $M_1,M_2\in\mathfrak{M}$ can be written as

$$M_i=\bigcup_{j=1}^{\infty}T_j^{\,i},\quad T_j^{\,i}\in\mathcal{T}(\mathcal{B}),\ i=1,2.$$

Then there exists $B_{ij}\in\mathcal{B}$ such that $B_{ij}\supset T_j^1$ and $B_{ij}\cap T_j^2=\emptyset$. Then $B\equiv\bigcup_{j=1}^{\infty}\bigcap_{i=1}^{\infty}B_{ij}\in\mathcal{B}$ satisfies $B\supset M_1$ and $B\cap M_2=\emptyset$.

"If" part. Take any θ_1 and θ_2 and define

$$q(x)=p(x,\theta_1)/\{p(x,\theta_1)+p(x,\theta_2)\}\quad\text{if}\quad p(x,\theta_1)+p(x,\theta_2)>0.$$

Take any $x\in[p(x,\theta_1)+p(x,\theta_2)>0]$ and consider $M\in\mathfrak{M}$ such that $x\in M$. Then, by Theorem 3.4, $M\subset[p(x,\theta_1)+p(x,\theta_2)>0]$ and q is constant on M. Therefore $[p(x,\theta_1)+p(x,\theta_2)>0]$ is a countable union of parts in $\mathfrak{M}$, say $=\bigcup_{i=1}^{\infty}M_i$. For any $i\neq j$ there exists $B_{ij}\in\mathcal{B}$ such that $B_{ij}\supset M_i$ and $B_{ij}\cap M_j=\emptyset$. Define $D_i=\cap\{B_{ij};j=1,\ldots,i-1,i+1,\ldots\}$. Then this sequence $\{D_i;i\geq 1\}$ is disjoint and satisfies $\bigcup_{i=1}^{\infty}D_i=\mathcal{X}$, $D_i\supset M_i$ for all i, and D_i meets no other M_j.

Of course D_i belongs to $\mathcal{B}$. Let us extend the function q to $\mathcal{X}$ in such a way that it is constant on each D_i. Then the extended function is $\mathcal{B}$-measurable and is the density of P_{θ_1} w.r.t. $P_{\theta_1}+P_{\theta_2}$, which implies $\mathcal{B}$ is pairwise sufficient by Lemma 2.1.

Take any M in $\mathfrak{M}$. Then $\sigma(M)$ denotes the subfield generated by all parts of $\mathfrak{M}$ except M.

Lemma 3.2. Suppose that $\mathcal{B}$ is a proper subfield of $\sigma(\mathfrak{M})$. Then $\mathcal{B}$ separates $\mathfrak{M}$ if and only if $\mathcal{B}=\sigma(M)$ for some $M\in\mathfrak{M}$.

Proof. Suppose that $\mathcal{B}$ separates $\mathfrak{M}$. Since $\mathcal{B}\neq\sigma(\mathfrak{M})$ there exists $M\in\mathfrak{M}$ such that $M\bar{\in}\mathcal{B}$. We show any $K\in\mathfrak{M}$ different from M belongs to $\mathcal{B}$. To prove this let us take such a part K. Then there exists $B\in\mathcal{B}$ such that $B\supset K$ and $B\cap M=\emptyset$ because $\mathcal{B}$ separates $\mathfrak{M}$. Since $B\in\sigma(\mathfrak{M})$ either B or $\mathcal{X}-B$ is a countable union of parts in $\mathfrak{M}$. In the latter case $\mathcal{X}-B$ has the form of

$$M\cup(\bigcup_{i=1}^{\infty}M_i),M\neq M_i,i=1,2,\ldots.$$

Then there exists $B_i\in\mathcal{B}$ such that $B_i\supset M$ and $B_i\cap M_i=\emptyset$. So we have

$$M=(\bigcap_{i=1}^{\infty}B_i)\cap(\mathcal{X}-B)\in\mathcal{B},$$

but this is a contradiction. Hence it must hold that

$$B=K\cup(\bigcup_{i=1}^{\infty}M_i),M_i\in\mathfrak{M}$$

for all i. Again by the separating property of $\mathcal{B}$ there exists, for each $i\geq1$, $B_i\in\mathcal{B}$ such that $B_i\supset K$ and $B_i\cap M_i=\emptyset$. Then we have $B\cap(\bigcup_{i=1}^{\infty}B_i)=K$, which implies that $K\in\mathcal{B}$. Hence we showed that $\sigma(M)\subset\mathcal{B}$. The converse inclusion is proved similarly, so we omit the detail. Hence we have $\mathcal{B}=\sigma(M)$.

If $\mathcal{B}=\sigma(M)$ holds then it is clear that $\mathcal{B}$ separates $\mathfrak{M}$. This completes the proof.

To state the final and main result in this section we need the following definition. A pairwise sufficient subfield

$\mathcal{B}$ is said to be *pairwise smallest sufficient* if, for any other pairwise sufficient subfield $\mathcal{C}$, it holds $\mathcal{B}\subset\tilde{\mathcal{C}}(\mathcal{A})$, where $\tilde{\mathcal{C}}(\mathcal{A})$ is the pairwise completion of $\mathcal{C}$ and is defined by

$$\{A\in\mathcal{A}; \text{for any } \theta_1,\theta_2 \text{ from } \Theta \text{ there exists } C\in\mathcal{C} \text{ such that } P_{\theta_i}(A\Delta C)=0,\ i=1,2\}.$$

The term "minimal pairwise sufficient" or "smallest pairwise sufficient" indicates the corresponding terminology about the ordering on the set of all pairwise sufficient subfields defined by "$\mathcal{B}$ is smaller than $\mathcal{C}$ if and only if $\mathcal{B}\subset\bar{\mathcal{C}}(\mathcal{A})$".

Theorem 3.8. (1) $\sigma(\mathfrak{M})$ is the smallest PSS subfield and is pairwise smallest sufficient.

(2) The subfields generated by all but any one of the parts in $\mathfrak{M}$ are minimal pairwise sufficient.

(3) There does not exist the smallest pairwise sufficient subfield.

Proof. (1) For any $\theta\in\Theta$ $S(P_\theta)\in\mathcal{B}(\mathfrak{M})$ and $S(P_\theta)$ is countable, so $S(P_\theta)\in\sigma(\mathfrak{M})$. By Theorem 3.5 $\sigma(\mathfrak{M})$ is pairwise sufficient. Hence $\sigma(\mathfrak{M})$ is PSS. Take any $\mathcal{B}$ which is PSS. By Theorem 3.6 $\mathcal{T}(\mathcal{B})>\mathcal{T}(\sigma(\mathfrak{M}))=\mathfrak{M}$. Then by Theorem 3.7 $\mathcal{B}\supset\sigma(\mathfrak{M})$. The latter half follows from Lemma 2.3 and 2.4.

(2) Take any $M\in\mathfrak{M}$. Since $\sigma(M)$ separates $\mathfrak{M}$ it is pairwise sufficient by Lemma 3.1. Take any pairwise sufficient subfield $\mathcal{B}$ for which it holds $\mathcal{B}\subset\sigma(M)$. By Lemma 3.1 $\mathcal{B}$ separates $\mathfrak{M}$. If $\mathcal{B}=\sigma(\mathfrak{M})$ then $\sigma(M)\subset\sigma(\mathfrak{M})=\mathcal{B}$. If $\mathcal{B}\neq\sigma(\mathfrak{M})$ then, by Lemma 3.2, there exists $N\in\mathfrak{M}$ such that $\sigma(N)=\mathcal{B}$. If $M\neq N$ then $M\in\mathcal{B}$. But then $M\in\sigma(M)$ by $\mathcal{B}\subset\sigma(M)$, and this is a contradiction. Hence $M=N$, which implies $\mathcal{B}=\sigma(M)$.

(3) Suppose that there exists the smallest pairwise sufficient subfield $\mathcal{C}$. Then $\mathcal{C}\subset\cap\{\sigma(M);M\in\mathfrak{M}\}$. Any M in $\mathfrak{M}$ does not belong to the right hand side. Hence $\mathcal{C}\subsetneq\sigma(\mathfrak{M})$. On the other hand, by Lemma 3.1, $\mathcal{C}$ separates $\mathfrak{M}$. Then there exists $N\in\mathfrak{M}$ such that $\mathcal{C}=\sigma(N)$ by Lemma 3.2. Of course there exists $M\in\mathfrak{M}$ such that $M\neq N$. This M belongs to $\sigma(N)$ and does not belong to $\mathcal{C}$. This is a contradiction. This completes the proof.

3.2. Majorized experiment case

In this section we consider majorized experiment $\mathcal{E}=$

$(\mathcal{X}, \mathcal{A}, \mathcal{P}=\{P_\theta; \theta\in\Theta\})$. Let $\mathcal{S}_0=\{S(P_\theta); \theta\in\Theta\}$ be the totality of supports, and let $\mathcal{S}$ be the σ-field generated by $\mathcal{S}_0$.

Lemma 3.3. There exists a maximal decomposition $\mathcal{F}$ of $\mathcal{E}$ such that $\mathcal{F}\subset\mathcal{S}$.

Proof. Clearly $\mathcal{E}(\mathcal{S})$ is a majorized experiment since $S(P_\theta)$ is a support of $P_\theta/\mathcal{S}$ w.r.t. $\mathcal{E}(\mathcal{S})$. Hence there exists a maximal decomposition $\mathcal{F}(\subset\mathcal{S})$ of $\mathcal{E}(\mathcal{S})$. We show that $\mathcal{F}$ is a maximal decomposition of $\mathcal{E}$. For this it only needs to show that for each $\theta\in\Theta$ $P_\theta(A)=0$ for all $A\in\mathcal{A}$ such that $P_\theta(A\cap F)=0$ for all $F\in\mathcal{F}$ and $\theta\in\Theta$. Take any $\theta\in\Theta$. Then P_θ concentrates on a countable union of sets in $\mathcal{F}$ because of the maximality of $\mathcal{F}$ for $\mathcal{E}(\mathcal{S})$. Then $P_\theta(A\cap F)=0$ for all F in this countable family, and hence $P_\theta(A)=0$.

Take any maximal decomposition $\mathcal{F}$ satisfying $\mathcal{F}\subset\mathcal{S}$ and let $F\in\mathcal{F}$. We define

$$\Theta(F)=\{\theta\in\Theta; P_\theta(F)>0\},$$

$$\mathcal{P}(F)=\{Q_\theta(\cdot)=P_\theta(\cdot)/P_\theta(F); \cdot\in\mathcal{A}\cap F, \theta\in\Theta(F)\},$$

and

$$\mathcal{E}(F)=(F, \mathcal{A}\cap F, \mathcal{P}(F)).$$

Then clearly $\mathcal{E}(F)$ is a dominated experiment. So there exists the smallest sufficient subfield, which we denote by $\mathcal{M}(F)$.

Lemma 3.4. Let $\mathcal{B}$ be a subfield of $\mathcal{A}$.
(1) If $\mathcal{B}$ is pairwise sufficient then so is $\mathcal{B}\cap F$ in $\mathcal{E}(F)$ for each F in $\mathcal{F}$.
(2) If $\mathcal{B}$ is the smallest PSS, then $\mathcal{B}\cap F$ is the smallest PSS for $\mathcal{E}(F)$ for each F in $\mathcal{F}$.

Proof. (1) Take any θ_1 and θ_2 from $\Theta(F)$. Then from Lemma 2.1, there exists a $\mathcal{B}$-measurable version of the density of P_{θ_1} w.r.t. $P_{\theta_1}+P_{\theta_2}$. Hence there exists a $\mathcal{B}$-measurable density of P_{θ_1} w.r.t. $P_{\theta_1}+(P_{\theta_1}(F)/P_{\theta_2}(F))P_{\theta_2}$ on F. But this version is also a version of the density of Q_{θ_1} w.r.t. $Q_{\theta_1}+Q_{\theta_2}$.

(2) Suppose that $\mathcal{B}$ is the smallest PSS. Then $\mathcal{B}\cap F$ is PSS for $\mathcal{E}(F)$ because of (1) and $S(P_\theta)\cap F$ is a support of Q_θ in $\mathcal{E}(F)$ for each $\theta\in\Theta(F)$. Take any PSS subfield $\mathcal{C}(F)$ for $\mathcal{E}(F)$. Define

$$\mathcal{C}=\{(B\cap(\mathcal{X}-F))\cup C; B\in\mathcal{B}, C\in\mathcal{C}(F)\}.$$

Then it is proved that $\mathcal{C}$ is PSS in $\mathcal{E}$. Hence $\mathcal{B}\subset\overline{\mathcal{C}}(\mathcal{A})$, and so $\mathcal{B}\cap F\subset\overline{\mathcal{C}[F]}(\mathcal{A}\cap F)$.

Next theorem gives a sufficient condition for pairwise sufficiency in terms of pairwise sufficiency for "restricted" statistical experiments.

Theorem 3.9. Let $\mathcal{B}$ be a subfield of $\mathcal{A}$ and $\mathcal{F}$ be a maximal decomposition of $\mathcal{E}$. If $\mathcal{B}\cap F$ is pairwise sufficient for $\mathcal{E}(F)$ for each $F\in\mathcal{F}$ and $\mathcal{B}$ separates $\mathcal{F}$, then $\mathcal{B}$ is pairwise sufficient.

Proof. Take any θ_1 and θ_2 from Θ. Then there exists a countable subfamily $\{F_1,F_2,...\}$ of $\mathcal{F}$ such that P_{θ_1} and P_{θ_2} concentrate on the union of sets in this subfamily. For each $i\geq1$ there exists a $\mathcal{B}\cap F_i$-measurable version of the density of Q_{θ_1} w.r.t. $Q_{\theta_1}+Q_{\theta_2}$. Then there exists a $\mathcal{B}$-measurable density f_i of P_{θ_1} w.r.t. $P_{\theta_1}+P_{\theta_2}$ on F_i. Since $\mathcal{B}$ separates $\mathcal{F}$ there exists a countable family $\{B_1,B_2,...\}$ from $\mathcal{B}$ such that $\mathcal{X}=\bigcup_{i=1}^{\infty}B_i$ and $F_i-B_i\in N(\mathcal{P})$ for all $i\geq1$. Let us extend f_i from F_i to B_i to get a $\mathcal{B}$-measurable function for each $i\geq1$, and combine them to get a function on $\mathcal{X}$. This is a $\mathcal{B}$-measurable version of the density of P_{θ_1} w.r.t. $P_{\theta_1}+P_{\theta_2}$.

Theorem 3.10. Let $\mathcal{F}$ be a maximal decomposition of $\mathcal{E}$ such that $\mathcal{F}\subset\bar{\mathcal{S}}(\mathcal{A})$. If $\mathcal{B}$ is pairwise sufficient for $\mathcal{E}$ then $\mathcal{B}$ separates $\mathcal{F}$.

Proof. Take any F_1 and F_2 from $\mathcal{F}$. Then there exists $G_i\in\mathcal{S}$ such that $P_\theta(F_i\Delta G_i)=0$ for all $\theta\in\Theta$, and $i=1,2$. Hence there exists a countable subfamily $\{P_{\theta_i};i\geq1\}$ of $\mathcal{P}$ such that G_1 and G_2 belongs to the subfield generated by $\{S(P_{\theta_i});i\geq1\}$, and $G_i\subset S(P_{\theta_i})$, $P_{\theta_i}(G_i)>0, i=1,2$.

Let $n=\sum_{i=1}^{\infty}2^{-i}P_{\theta_i}$, and let $\mathcal{P}_0=\{n,P_{\theta_1},P_{\theta_2},...\}$. Consider experiment $\mathcal{E}_0=(\mathcal{X},\mathcal{A},\mathcal{P}_0)$. Then $\mathcal{X}$ and $S(P_{\theta_i})$ are supports for $\mathcal{E}_0$ of n and P_{θ_i} respectively. Let $\mathcal{S}_0$ be the subfield generated by these supports. Then $\mathcal{F}_0=\{G_1,G_2,\mathcal{X}-(G_1\cup G_2)\}$ is a maximal decomposition of $\mathcal{E}_0$ such that $\mathcal{F}_0\subset\mathcal{S}_0$. Since $\mathcal{B}$ is pairwise sufficient it is sufficient for the dominated experiment $\mathcal{E}_0$. Hence $\mathcal{B}$ separates $\mathcal{F}_0$. Then there exists $B\in\mathcal{B}$ such that $G_1-B \in N(\mathcal{P}_0)$ and $G_2\cap B\in N(\mathcal{P}_0)$. This shows $\mathcal{B}$ separates $\{F_1,F_2\}$.

Next theorem gives a sufficient condition for PSS in terms of "restricted" experiments. In the proof of the following theorem we denote $A\in\mathcal{B}[\mathcal{P}]$ if there exists a set $B\in\mathcal{B}$ such that $A\Delta B\in N(\mathcal{P})$, and $A=B$ $[\mathcal{P}]$ if $A\Delta B\in N(\mathcal{P})$.

Theorem 3.11. Let $\mathcal{B}$ be a subfield of $\mathcal{A}$ and $\mathcal{F}$ be a maximal decomposition of $\mathcal{E}$. If $\mathcal{B}_\cap F$ is pairwise sufficient for $\mathcal{E}(F)$ for each $F\in\mathcal{F}$ and if $\mathcal{F}\subset\bar{\mathcal{B}}(\mathcal{A})$, then $\mathcal{B}$ is PSS.

Proof. By assumption $\bar{\mathcal{B}}(\mathcal{A})\supset\mathcal{F}$, so $\mathcal{B}$ separates $\mathcal{F}$. Then, by Theorem 3.9, $\mathcal{B}$ is pairwise sufficient. Since $\mathcal{B}_\cap F$ is pairwise sufficient for the dominated experiment $\mathcal{E}(F)$, we have $\mathcal{M}(F)\subset \overline{\mathcal{B}_\cap F}(\mathcal{A})$. Then, by the assumption $\bar{\mathcal{B}}(\mathcal{A})\supset\mathcal{F}$, it follows that $\mathcal{M}(F) \subset\bar{\mathcal{B}}(\mathcal{A})$. Take any $\theta\in\Theta$ and $F\in\mathcal{F}$ such that $P_\theta(F)>0$. Then $S(P_\theta)\cap F$ is a support for $\mathcal{E}(F)$ of Q_θ and, as any sufficient subfield in a dominated experiment contains supports, $S(P_\theta)\cap F\in\mathcal{M}(F)$ $[\mathcal{P}(F)]$. Hence $S(P_\theta)\cap F\in\mathcal{B}[\mathcal{P}]$. For each $\theta\in\Theta$ there exists a countable family $\{F_1,F_2,\ldots\}$ on the union of which P_θ is concentrated. Hence

$$S(P_\theta)=\bigcup_{i=1}^{\infty}(S(P_\theta)\cap F_i)\ [\mathcal{P}],$$

which implies $S(P_\theta)\in\mathcal{B}[\mathcal{P}]$ because $S(P_\theta)\cap F_i\in\mathcal{B}[\mathcal{P}]$ for all $i\geq 1$.

Theorem 3.12. Let $\mathcal{F}$ be a maximal decomposition of $\mathcal{E}$ such that $\mathcal{F}\subset\bar{\mathcal{J}}(\mathcal{A})$ and let $\mathcal{B}$ be a subfield of $\mathcal{A}$. If $\mathcal{B}$ is PSS then $\mathcal{F}\subset\bar{\mathcal{B}}(\mathcal{A})$.

Proof. By the definition of $\mathcal{J}$ we have $\mathcal{J}\subset\bar{\mathcal{B}}(\mathcal{A})$, which implies $\mathcal{F}\subset\bar{\mathcal{B}}(\mathcal{A})$.

Here we summarize the main results of this section which characterize the structure of pairwise sufficiency, PSS and the smallest PSS extending the corresponding results in discrete experiment.

Theorem 3.13. Take any maximal decomposition $\mathcal{F}$ such that $\mathcal{F}\subset\bar{\mathcal{J}}(\mathcal{A})$ and suppose that $\mathcal{B}\cap F$ is pairwise sufficient for $\mathcal{E}(F)$ for each $F\in\mathcal{F}$. Then $\mathcal{B}$ is pairwise sufficient if and only if $\mathcal{B}$ separates $\mathcal{F}$, and $\mathcal{B}$ is PSS if and only if $\bar{\mathcal{B}}(\mathcal{A})\supset\mathcal{F}$.

Theorem 3.14. Let $\mathcal{F}$ be any maximal decomposition such that $\mathcal{F}\subset\bar{\mathcal{J}}(\mathcal{A})$. The following σ-field is the smallest PSS subfield:

$\{A\in\mathcal{A}$; (1) $A\cap F$ is an element of the smallest sufficient subfield for the dominated experiment $\mathcal{E}(F)$ for each $F\in\mathcal{F}$, and

(2-1) $A\cap F=\emptyset\,[\mathcal{P}]$ holds for all but a countable number of F in $\mathcal{F}$,

or

(2-2) $A\cap F=F\,[\mathcal{P}]$ holds for all but a countable number of F in $\mathcal{F}\}$.

Theorem 3.13 follows from Theorem 3.11 and 3.12. Theorem 3.14 follows from Theorem 3.13.

We note that the construction of the smallest PSS subfield for majorized experiment can be achieved using simply the smallest sufficient subfields for the "restricted" dominated experiments which are generated by a maximal decomposition of the original experiment.

As for the structure of pairwise smallest sufficient subfield the next resut will give information.

Theorem 3.15. Let $\mathcal{E}$ be a majorized experiment. Then the smallest PSS subfield is pairwise smallest sufficient.

Proof. Let $\mathcal{B}$ be the smallest PSS for $\mathcal{E}$. Take any pairwise sufficient subfield $\mathcal{C}$. Suppose that $\mathcal{D}$ is the subfield generated by $\mathcal{C}$ and $\mathcal{S}_0$. We show $\mathcal{D}\subset\tilde{\mathcal{C}}(\mathcal{A})$. For this purpose it is enough to show $S(P_\theta)\in\tilde{\mathcal{C}}(A)$ for all $\theta\in\Theta$. For any θ_1, θ_2 from Θ there exists a $\mathcal{C}$-measurable density f of P_θ w.r.t. $P_{\theta_1}+P_{\theta_2}+P_\theta$ because $\mathcal{C}$ is pairwise sufficient. Then we have, for $C=[f(x)>0]$,

$$\int_{C\cap(\mathcal{X}-S(P_\theta))} f(x)\ d(P_{\theta_1}+P_{\theta_2}+P_\theta)=P_\theta(C\cap(\mathcal{X}-S(P_\theta)))=0.$$

Hence $(P_{\theta_1}+P_{\theta_2})(C\cap(\mathcal{X}-S(P_\theta)))=0$. On the other hand

$$P_\theta((\mathcal{X}-C)\cap S(P_\theta))\leqq P_\theta(\mathcal{X}-C)=0.$$

Thus

$$(P_{\theta_1}+P_{\theta_2})(C\Delta S(P_\theta))=0.$$

This implies $S(P_\theta)\in\tilde{\mathcal{C}}(\mathcal{A})$, so that $\mathcal{D}\subset\tilde{\mathcal{C}}(\mathcal{A})$. Since $\mathcal{D}$ is PSS we have $\mathcal{B}\subset\bar{\mathcal{D}}(\mathcal{A})$. Then it is clear that $\bar{\mathcal{D}}(\mathcal{A})\subset\tilde{\mathcal{C}}(\mathcal{A})$. Hence it follows that $\mathcal{B}$ is pairwise smallest sufficient.

We note that, by Theorem 3.8, the converse of Theorem 3.15 does not hold.

3.3. Burkholder problem of sufficiency and completions

In order to prove Theorem 3.1 using "general" results we need, in addition to Theorems in Chapter 2, a few notions of completions of subfields. Then these are applied to the problem of order preservation of sufficiency, i.e. the Burkholder problem.

Let $\mathcal{E}=(\mathcal{X},\mathcal{A},\mathcal{P}=\{P_\theta;\theta\in\Theta\})$ be any experiment and let $\mathcal{B}$ be a subfield of $\mathcal{A}$. Then we define weak self-completion

$\hat{\mathcal{B}}(\mathcal{B})=\{A\in\mathcal{A}$; for any $\theta\in\Theta$ there exist sets $B\in\mathcal{B}$ and $N\in\mathcal{B}$ such that $A\Delta B\subset N[\mathcal{P}]$ and $P_\theta(N)=0\}$, and

pairwise self-completion

$\tilde{\mathcal{B}}(\mathcal{B})=\{A\in\mathcal{A}$; for any $\{\theta_1,\theta_2\}\subset\Theta$ there exist $B\in\mathcal{B}$, $N\in\mathcal{B}$ such that $A\Delta B\subset N[\mathcal{P}]$ and $P_{\theta_i}(N)=0$ for $i=1,2\}$. Here $A\subset B[\mathcal{P}]$ means $A-B\in N(\mathcal{P})$.

Lemma 3.5. Let $\mathcal{E}=(\mathcal{X},\mathcal{A},\mathcal{P}=\{P_\theta;\theta\in\Theta\})$ be any experiment and let $\mathcal{B}$ be pairwise sufficient. Then $\hat{\mathcal{B}}(\mathcal{B})=\tilde{\mathcal{B}}(\mathcal{B})$.

Proof. Take any $A\in\hat{\mathcal{B}}(\mathcal{B})$ and P_{θ_1} and P_{θ_2} from $\mathcal{P}$. Then there exist $B_i\in\mathcal{B}$ and $N_i\in\mathcal{B}$ such that $A\Delta B_i\subset N_i[\mathcal{P}]$ and $P_{\theta_i}(N_i)=0$ for $i=1,2$. Because of sufficiency of $\mathcal{B}$ for $\{P_{\theta_1},P_{\theta_2}\}$, there exists a $\mathcal{B}$-measurable function h such that

$$\int_B h(x)\ dP_{\theta_i}=P_{\theta_i}(A\cap B)=P_{\theta_i}(A\cap B_{\theta_i})$$

for all $B\in\mathcal{B}$ and $i\in 1,2$. So we have $P_{\theta_i}(B_i\Delta B_{1,2})=0$ for $i=1,2$, where $B_{1,2}=[h(x)=1]\in\mathcal{B}$. Next we show

$$(A\Delta B_{1,2})-\bigcap_{i=1}^{2}\{(B_{1,2}\Delta B_i)\cup N_i\} \tag{3.2}$$

is a $\mathcal{P}$-null set. By

$$(A-B_{1,2})-\{(B_{1,2}\Delta B_1)\cup N_i\}\subset(A-B_1)-N_1\subset(A\Delta B_1)-N_1$$
and
$$(B_{1,2}-A)-\{(B_{1,2}\Delta B_1)\cup N_1\}\subset(B_1-A)-N_1\subset(A\Delta B_1)-N_1,$$

we have

$$(A\Delta B_{1,2})-\{(B_{1,2}\Delta B_1)\cup N_1\}\subset(A\Delta B_1)-N_1,$$

and $(A\Delta B_1)-N_1$ is a $\mathcal{P}$-null set. Similarly it follows that

$$(A\Delta B_{1,2})-\{(B_{1,2}\Delta B_2)\cup N_2\}\subset(A\Delta B_2)-N_2$$

and $(A\Delta B_2)-N_2$ is a $\mathcal{P}$-null set. Hence the set in (3.2) is a $\mathcal{P}$-null set. If we denote $N_{1,2}=\bigcap_{i=1}^{2}\{(B_{1,2}\Delta B_i)\cup N_i\}$, $N_{1,2}\in\mathcal{B}$, $A\Delta B_{1,2}\subset N_{1,2}\,[\mathcal{P}]$ and $P_{\theta_i}(N_{1,2})=0$ for $i=1,2$. Hence we have $A\in\tilde{\mathcal{B}}(\mathcal{B})$, which implies $\hat{\mathcal{B}}(\mathcal{B})\subset\tilde{\mathcal{B}}(\mathcal{B})$. The converse is obvious.

<u>Theorem 3.16</u>. Let $\mathcal{E}$ be weakly dominated and let $\mathcal{B}$ be a subfield. Then $\mathcal{B}$ is PSS if and only if $\tilde{\mathcal{B}}(\mathcal{B})$ is sufficient.

<u>Proof</u>. "Only if" part. For any $A\in\hat{\mathcal{B}}(\mathcal{A})$ and $\theta\in\Theta$ there exist sets $B_\theta\in\mathcal{B}$ and $N_\theta\in\mathcal{A}$ such that $A\Delta B_\theta\subset N_\theta$ and $P_\theta(N_\theta)=0$. By the assumption on supports there exists a version $p(x,\theta)$ of the density of P_θ w.r.t. an equivalent localizable majorizing measure m such that $[p(x,\theta)>0]\in\mathcal{B}$. Then by $m\sim\mathcal{P}$ we have

$$N_\theta\subset\mathcal{X}-[p(x,\theta)>0]\quad[\mathcal{P}].$$

Hence we have $A\Delta B_\theta\subset\mathcal{X}-[p(x,\theta)>0]\ [\mathcal{P}]$. This implies $A\in\hat{\mathcal{B}}(\mathcal{B})$. So we have $\hat{\mathcal{B}}(\mathcal{A})=\hat{\mathcal{B}}(\mathcal{B})$ because $\hat{\mathcal{B}}(\mathcal{B})\subset\hat{\mathcal{B}}(\mathcal{A})$ is obvious. By Lemma 3.5

$\hat{\mathcal{B}}(\mathcal{B})=\tilde{\mathcal{B}}(\mathcal{B})$. On the other hand $\hat{\mathcal{B}}(\mathcal{A})$ is sufficient by Lemma 2.3. Thus $\tilde{\mathcal{B}}(\mathcal{B})$ is sufficient.

"If" part. For any $A\in\mathcal{A}$ there exists a $\tilde{\mathcal{B}}(\mathcal{B})$-measurable function g such that

$$\int_B g(x)\ dP_\theta = P_\theta(A\cap B) \text{ for all } \tilde{B}(\mathcal{B}) \text{ and } \theta\in\Theta.$$

Take any pair $\{P_{\theta_1},P_{\theta_2}\}$ from $\mathcal{P}$. Then there exists a $\mathcal{B}$-measurable function h such that $g(x)=h(x)$ a.e.P_{θ_i}, $i=1,2$. Then we have

$$\int_B h(x)\ dP_{\theta_i}=\int_B g(x)\ dP_{\theta_i}=P_{\theta_i}(A\cap B) \text{ for all } B\in\mathcal{B} \text{ and } i=1,2.$$

So $\mathcal{B}$ is pairwise sufficient. For any $\theta\in\Theta$ there exists a version $p(x,\theta)$ of the density of P_θ w.r.t. an equivalent majorizing measure m such that $A(\theta)\equiv[p(x,\theta)>0]\in\tilde{\mathcal{B}}(\mathcal{B})$ because $\tilde{\mathcal{B}}(\mathcal{B})$ is sufficient. Then there exist sets $B_\theta\in\mathcal{B}$ and $N_\theta\in\mathcal{B}$ such that $A(\theta)\Delta B_\theta\subset N_\theta[\mathcal{P}]$ and $P_\theta(N_\theta)=0$. Define

$$M(\theta)=(A(\theta)\Delta B_\theta)-N_\theta,\ B_\theta'=B_\theta-N_\theta \text{ and } M_\theta'=(A(\theta)\Delta B_\theta')-N_\theta.$$

Then it follows that $A(\theta)\Delta B_\theta'\subset N_\theta\ [\mathcal{P}]$ because $M_\theta'\subset M(\theta)$. Further we have $B_\theta'-A(\theta)\subset M_\theta'$ because $B_\theta'-M_\theta'\subset A(\theta)-M_\theta'$. Define a function $q(\cdot,\theta)$ by

$$\begin{aligned} q(x,\theta)&=p(x,\theta) && \text{if } x\in B_\theta'\cap A(\theta),\\ &=1 && \text{if } x\in B_\theta'-A(\theta),\\ &=0 && \text{otherwise.}\end{aligned}$$

Then it follows that $[q(x,\theta)>0]=B_\theta'\in\mathcal{B}$ and $[p(x,\theta)\neq q(x,\theta)]=A_\theta'\Delta B_\theta'$. But $m(A_\theta'\Delta B_\theta')=0$ because m is an equivalent majorizing measure. Hence $q(\cdot,\theta)$ is also a version of the density. And $\mathcal{B}$ contains a support of P_θ for all $\theta\in\Theta$. This completes the proof.

<u>Theorem 3.17</u>. Suppose that $\mathcal{E}=(\mathcal{X},\mathcal{A},\mathcal{P}=\{P_\theta;\theta\in\Theta\})$ be a discrete experiment. If a subfield $\mathcal{B}$ is closed then $\tilde{\mathcal{B}}(\mathcal{B})=\mathcal{B}$. If $\mathcal{B}$ is PSS then the converse is also true.

Proof. Take any $A \in \tilde{\mathcal{B}}(\mathcal{B})$. For any $\theta \in \Theta$ there exist B_θ, N_θ in $\mathcal{B}$ such that $A \Delta B_\theta \subset N_\theta$, $P_\theta(N_\theta)=0$. Note that we can assume $B \subset A$ without loss of generality, so $A-B_\theta \subset N_\theta$. Let's define $B=\cup\{B_\theta;\theta\in\Theta\}$, $N=\cap\{N_\theta;\theta\in\Theta\}$. Then B and N belong to $\mathcal{B}$ by the assumption, and satisfy $A-B\subset N$. Since $N\in N(\mathcal{P})$, $N=\emptyset$. This shows $A=B\in\mathcal{B}$.

Conversely if $\mathcal{B}$ is PSS and $\tilde{\mathcal{B}}(\mathcal{B})=\mathcal{B}$ then, by Theorem 3.16, $\mathcal{B}$ is sufficient. By Theorem 3.2 and Proposition 3.1 $\mathcal{B}$ is closed.

Now let's give another proof of Theorem 3.1. If $\mathcal{T}$ is sufficient, that is $\mathcal{B}(\mathcal{T})$ is sufficient, by Theorem 2.5 we have

$$p(x,\theta)=g(x,\theta)\ h(x) \tag{3.3}$$

for all $x\in\mathcal{X},\theta\in\Theta$. Here $g(\cdot,\theta)$ is $\mathcal{B}(\mathcal{T})$-measurable, so it is constant on each part of $\mathcal{T}$. Then by (3.3), it follows that

$$p(x,\theta)=P_\theta(T_x)h'(x)$$

for all $x\in\mathcal{X},\theta\in\Theta$.

Conversely suppose that (3.3) holds for each $x\in\mathcal{X},\theta\in\Theta$. Since $P_\theta(T_x)$, as a function of x, is constant on each part of $\mathcal{T}$, it is $\mathcal{B}(\mathcal{T})$-measurable. By Theorem 2.5 $\mathcal{B}(\mathcal{T})$ is PSS. If we define $\mathcal{C}=\mathcal{B}(\mathcal{T})$ then $\tilde{\mathcal{C}}(\mathcal{C})$ is sufficient by Theorem 3.16. But, since $\mathcal{C}$ is closed, $\tilde{\mathcal{C}}(\mathcal{C})=\mathcal{C}$ by Theorem 3.17. This implies $\mathcal{C}$ is sufficient. Thus $\mathcal{T}$ is sufficient by the definition.

Next we apply the completions to the Burkholder problem.

Theorem 3.18. Let $\mathcal{E}$ be a discrete experiment. Suppose that $\mathcal{T}_1>\mathcal{T}_2$ and $\mathcal{T}_2$ is sufficient. Then $\mathcal{T}_1$ is also sufficient.

Proof. By assumptions $\mathcal{B}(\mathcal{T}_1)\supset\mathcal{B}(\mathcal{T}_2)$ and $\mathcal{B}(\mathcal{T}_2)$ is sufficient. Since $\mathcal{B}(\mathcal{T}_1)$ is closed we have $\tilde{\mathcal{C}}(\mathcal{C})=\mathcal{C}$ by Theorem 3.17, where $\mathcal{C}=\mathcal{B}(\mathcal{T}_1)$. On the other hand, since $\mathcal{C}$ is PSS, $\hat{\mathcal{C}}(\mathcal{C})$ is sufficient by Theorem 3.16. Hence $\mathcal{T}_1$ is sufficient.

Theorem 3.18 can also be proved from Theorem 3.1.
For any subfield $\mathcal{B}$ we define

$\bar{\bar{\mathcal{B}}}(\mathcal{B})=\{A\in\mathcal{A};$ there exist $B\in\mathcal{B}$ and $N_\theta\in\mathcal{B}$ for all $\theta\in\Theta$ such that $A\Delta B\subset N_\theta[\mathcal{P}]$ and $P_\theta(N_\theta)=0\}$.

Theorem 3.19. Let $\mathcal{E}=(\mathcal{X},\mathcal{A},\mathcal{P}=\{P_\theta;\theta\in\Theta\})$ be weakly dominated. Let $\mathcal{B}$ be PSS and $\mathcal{B}\subset\mathcal{C}$. Then $\mathcal{C}$ is sufficient if and only if $\hat{\mathcal{C}}(\mathcal{C})=\bar{\bar{\mathcal{C}}}(\mathcal{C})$.

Proof. Suppose that $\mathcal{C}$ is sufficient. For any $A\in\hat{\mathcal{C}}(\mathcal{C})$ and $\theta\in\Theta$ there exist $C_\theta,N_\theta\in\mathcal{C}$ such that $A\Delta C_\theta\subset N_\theta[\mathcal{P}],P_\theta(N_\theta)=0$. Let's define $f(x,\theta)=I_{C_\theta}(x)$. Then $\{f(\cdot,\theta);\theta\in\Theta\}$ is a bounded finitely coherent family of $\mathcal{C}$-measurable functions. This follows from, since $\mathcal{C}$ is pairwise sufficient, $\hat{\mathcal{C}}(\mathcal{C})=\{A\in\mathcal{A};$ for any finite $\{\theta_1,\theta_2,\ldots,\theta_k\}\subset\Theta$ there exist $C,N\in\mathcal{C}$ such that $A\Delta C\subset N[\mathcal{P}],P_{\theta_i}(N)=0$ $,i=1,2,\ldots,k\}$. On the other hand, by Lemma 2.5, $\mathcal{E}(\mathcal{C})$ is weakly dominated. Then, by Theorem 1.5, there exists a $\mathcal{C}$-measurable bounded function f such that $f(x)=f(x,\theta)$ a.e. P_θ for all $\theta\in\Theta$. If we define $C=[f(x)=1]$, we have, $P_\theta(C\Delta C_\theta)=0$ for all $\theta\in\Theta$. $A\Delta C\subset N_\theta\cup(C_\theta\Delta C)\,[\mathcal{P}]$ and $P_\theta(N_\theta\cup(C_\theta\Delta C))=0$ imply that $A\in\bar{\bar{\mathcal{C}}}(\mathcal{C})$. Thus we proved $\hat{\mathcal{C}}(\mathcal{C})\subset\bar{\bar{\mathcal{C}}}(\mathcal{C})$. The converse inclusion is obvious.

Conversely let's assume that $\hat{\mathcal{C}}(\mathcal{C})=\bar{\bar{\mathcal{C}}}(\mathcal{C})$. Since $\mathcal{C}$ is PSS $\hat{\mathcal{C}}(\mathcal{C})$ is sufficient by Theorem 3.16. Thus $\bar{\bar{\mathcal{C}}}(\mathcal{C})$ is sufficient. Then it is easy to show that $\mathcal{C}$ is sufficient.

Corollary 3.2. Let $\mathcal{E}$ be a discrete experiment. Let $\mathcal{B}$ be PSS and $\mathcal{B}\subset\mathcal{C}$. Then $\mathcal{C}$ is sufficient if and only if $\mathcal{C}=\hat{\mathcal{C}}(\mathcal{C})$, and this condition is also necessary and sufficient for $\mathcal{C}$ to be closed.

Proof. If $\mathcal{C}$ is sufficient we have $\bar{\bar{\mathcal{C}}}(\mathcal{C})=\mathcal{C}$ because $\mathcal{C}$ is closed and $N(\mathcal{P})=\{\emptyset\}$. So we have $\mathcal{C}=\hat{\mathcal{C}}(\mathcal{C})$ by Theorem 3.19. The latter half of the Corollary follows from Corollary 3.1, Theorem 3.16 and Theorem 3.17.

3.4. Bibliographical notes

Basu and Ghosh(1967) introduced the discrete experiment as a mathematical model of sampling from a finite universe. They constructed, in particular, the smallest sufficient statistic. This experiment was also studied in detail by

Morimoto(1972,1973), Ghosh, Morimoto and Yamada(1981). Section 3.2 is exclusively indebted to Fujii and Morimoto(1986). Fujii (1989,1991) gave set-theoretical conditions of the existence of the smallest pairwise sufficient subfield. The results of Section 3.3. were taken from Yamada (1980,1993). Other characterizations, related to L- and M-space, of pairwise sufficiency are found in Torgersen(1991). The book of Torgersen(1991) also contains results on discrete experiments which are free from the condition $N(\mathcal{P})=\{\emptyset\}$.

In Section 3.3, we observed that the notions "closed" and "completion" are conditions to avoid the Burkholder's pathological examples. In his Section 3 Morimoto(1972) asserted, based upon his and Basu and Ghosh(1967) results, to treat rather "complete" σ-fields (that is, closed σ-fields) and "completely additive" probability measures when we consider a statistical model of sampling from a finite population. By replacing this "complete" sigma field with completions of the Boolean algebra Michaletzky(1985) studied the condition to avoid the Burkholder phenomena. There Michaletzky considers an experiment $(\mathcal{X},\mathcal{A},\mathcal{P})$ in which the Boolean algebra $\mathcal{A}/N(\mathcal{P})$ is complete, and elements of $\mathcal{P}$ are not restricted to discrete probability measures. Suppose that $\mathcal{B}$ is sufficient, $N(\mathcal{P})\subset\mathcal{B}$ and $\mathcal{B}\subset\mathcal{C}$. Then $\mathcal{C}$ is sufficient if and only if the corresponding Boolean algebra of $\mathcal{C}$ is a complete Boolean subalgebra of $\mathcal{A}/N(\mathcal{P})$. For more detailed discussion we refer to Michaletzky (1987), Fujii and Morimoto (1993).

Pairwise smallest sufficient subfield in a majorized experiment was constructed by Luschgy(1978), Siebert(1979*a*) and Ghosh,Morimoto and Yamada(1981) independently. Luschgy (1978) and Siebert (1979*a*) gave examples of non-majorized experiments for which there do not exist pairwise smallest sufficient subfields.

Roy and Ramamoorthi (1979), Ramamoorthi (1980,1982) includes studies of relationship among Bayes sufficiency, sufficiency, and pairwise sufficiency especially when the σ-fields are countably generated. Blackwell and Ramamoorthi (1982) pointed out, in a standard Borel space, that there exists a Bayes sufficient statistic which is not sufficient. For related works we refer to Martin,Petit and Littaye (1973), Yamada and Suzuki (1990).

Chapter 4. The Rao-Blackwell theorem and UMVUE

The Rao-Blackwell theorem is one of the most important theorems in mathematical statistics. It asserts that any unbiased estimator is improved w.r.t. variance by an unbiased estimator which is a function of a sufficient statistic. Hence the class of unbiased estimators which are functions of a sufficient statistic constitutes an essentially complete class. However such a "Rao-Blackwellization" of an unbiased estimator does not necessarily provide a UMVU (uniformly minimum variance unbiased) estimator. Lehmann and Scheffé (1950) showed that if there exists a complete sufficient statistic then the Rao-Blackwellization w.r.t. the statistic produces a UMVU estimator.

The aim of this Chapter is to answer the following questions:

(1) If any unbiased estimator is improved by an unbiased estimator which is a function of a fixed statistic, then is the statistic sufficient?

(2) If there exists a UMVU estimator for any unbiasedly estimable parameter function, necessarily does there exist a complete sufficient statistic?

This Chapter is mainly indebted to Lehmann and Scheffé (1950), Bahadur(1957), Takeuchi(1969), Yamada(1980) and Torgersen(1988).

4.1. Rao-Blackwell theorem for PSS in weakly dominated experiments

Definition 4.1. Let $\mathcal{E}=(\mathcal{X},\mathcal{A},\mathcal{P}=\{P_\theta;\theta\in\Theta\})$ be any experiment and let $\mathcal{B}$ be a subfield of $\mathcal{A}$. Then $\mathcal{B}$ is said to have the *Rao-Blackwell property* if the following property w.r.t. $\mathcal{B}$ holds: for any $t\in\cap\{L^2(\mathcal{X},\mathcal{A},P_\theta);\theta\in\Theta\}$ there exists a $\mathcal{B}$-measurable function g such that

$$\int t(x)\ dP_\theta = \int g(x)\ dP_\theta, \qquad \text{and}$$

$$\int t^2(x)\ dP_\theta \geq \int g^2(x)\ dP_\theta \text{ for all } \theta\in\Theta.$$

Here $L^2(\mathcal{X},\mathcal{A},P_\theta)$ is the totality of (equivalence classes of) functions f for which f^2 is integrable w.r.t. P_θ.

An answer to the question (1) above was given by Takeuchi (1969) without any restriction.

Theorem 4.1. If $\mathcal{B}$ has the Rao-Blackwell property then it is pairwise sufficient.

Proof. Take any θ_1 and θ_2 from Θ and put

$$t(x)=dP_{\theta_1}/d(P_{\theta_1}+P_{\theta_2})(x).$$

Then by the assumption there exists a $\mathcal{B}$-measurable function g such that

$$\int t(x)\ dP_\theta = \int g(x)\ dP_\theta, \qquad \text{and}$$

$$\int t^2(x)\,dP_\theta \geq \int g^2(x)\ dP_\theta$$

for all θ. Then it follows that

$$\int t^2(x)\ d(P_{\theta_1}+P_{\theta_2}) = \int t(x)\ dP_{\theta_1} = \int g(x)\ dP_{\theta_1} =$$

$$\int t(x)g(x)\ d(P_{\theta_1}+P_{\theta_2}).$$

So we have

$$0 \leqq \int (t(x) - g(x))^2\ d(P_{\theta_1}+P_{\theta_2})$$

$$= \int g^2(x)\ d(P_{\theta_1}+P_{\theta_2}) - \int t^2(x)\ d(P_{\theta_1}+P_{\theta_2}) \leqq 0$$

by the Rao-Blackwell property. Therefore $t(x)=g(x)$ a.e. $P_{\theta_1}+P_{\theta_2}$ holds. Hence g is a $\mathcal{B}$-measurable version of the density of P_{θ_1} w.r.t. $P_{\theta_1}+P_{\theta_2}$, which implies that $\mathcal{B}$ is pairwise sufficient by Lemma 2.1 because θ_1 and θ_2 are arbitrary.

The next theorem is usually referred to as the Rao-Blackwell theorem.

Theorem 4.2. If $\mathcal{B}$ is sufficient then it has the Rao-Blackwell property.

Proof. Take any $t \in \cap\{L^2(\mathcal{X},\mathcal{A},P_\theta);\theta\in\Theta\}$. Then $g(x)=E(t/\mathcal{B})(x)$

satisfies the conditions of the Rao-Blackwell property by Jensen's inequality.

Corollary 4.1. Suppose that $\mathscr{E}$ is dominated. Then a subfield $\mathscr{B}$ has the Rao-Blackwell property if and only if $\mathscr{B}$ is sufficient.

This follows from Theorem 4.1 and Theorem 4.2.

The converse of Theorem 4.2 does not hold for an undominated experiment. This is due to the Burkholder phenomena. Take subfields $\mathscr{B}_1$ and $\mathscr{B}_2$ such that $\mathscr{B}_1 \subset \mathscr{B}_2$, and such that $\mathscr{B}_1$ is sufficient but $\mathscr{B}_2$ is not. Then by Theorem 4.2 $\mathscr{B}_1$ and hence $\mathscr{B}_2$ has the Rao-Blackwell property. Note that the Burkholder phenomena may occur in the case of weak domination (Section 3.1).

We will study here the relationship between PSS and the Rao-Blackwell property.

Lemma 4.1. Let $\mathscr{E}$ be weakly dominated and $\mathscr{B}$ be a subfield of $\mathscr{A}$. If $\mathscr{B}$ contains supports for $\mathscr{E}$ and $\hat{\mathscr{B}}(\mathscr{A})$ is sufficient, then $\mathscr{B}$ is pairwise sufficient.

Proof. Let m be an equivalent localizable majorizing measure for $\mathscr{E}$. By Theorem 2.4 we have the Neyman factorization

$$dP_\theta/dm(x) = g(x,\theta)h(x) \quad \text{a.e. } m,$$

where $g(\cdot,\theta)$ is a non-negative $\hat{\mathscr{B}}(\mathscr{A})$-measurable function and h is a non-negative $\mathscr{A}$-measurable function. Hence, for each $\theta \in \Theta$, there exists a $\mathscr{B}$-measurable function $g'(\cdot,\theta)$ such that $g(x,\theta) = g'(x,\theta)$ a.e. P_θ. So we have

$$dP_\theta/dm(x) = g'(x,\theta)h(x) \quad a.e.\ P_\theta.$$

Since $\mathscr{B}$ contains supports and m is an equivalent majorizing measure, for each $\theta \in \Theta$, there exists a version $p(\cdot,\theta)$ of the density of P_θ w.r.t. m such that $[p(x,\theta)>0] \in \mathscr{B}$. Let us define $g''(x,\theta)$ by

$$g''(x,\theta) = g'(x,\theta) \qquad \text{if } p(x,\theta)>0,$$

$$= 0 \qquad\qquad \text{otherwise,}$$

and denote

$$N(\theta)=[p(x,\theta)\neq g'(x,\theta)\ h(x)].$$

Since

$$0=P_\theta(N(\theta))=P_\theta(N(\theta)\cap[p(x,\theta)>0])$$

$$=\int_{N(\theta)\cap[p(x,\theta)>0]} p(x,\theta)\ dm(x),$$

we have

$$m(N(\theta)\cap[p(x,\theta)>0])=0.$$

But then it follows that

$$p(x,\theta)=g''(x,\theta)h(x) \text{ a.e. } m$$

for each $\theta\in\Theta$. Since $g''(\cdot,\theta)$ is non-negative and $\mathcal{B}$-measurable, $\mathcal{B}$ is PSS by Theorem 2.4. In particular $\mathcal{B}$ is pairwise sufficient. This completes the proof.

Remark 4.1. In Lemma 4.1 we can not drop the assumption that $\mathcal{B}$ contains supports for $\mathcal{E}$. For example let $\mathcal{X}$ be the real line, and $\mathcal{A}$ be the power set of $\mathcal{X}$. Let $\mathcal{P}=\{P_x;x\in\mathcal{X}\}$ be the totality of point probability measures on $(\mathcal{X},\mathcal{A})$. Let $\mathcal{B}$ be the subfield generated by all singletons but 0 and 1, that is $\mathcal{B}=\{A\in\mathcal{A};A$ is countable and does not contain 0 and 1, or $\mathcal{X}-A$ is countable and contains 0 and 1$\}$. Then it can be proved that $\hat{\mathcal{B}}(\mathcal{A})=\mathcal{A}$, which shows that $\hat{\mathcal{B}}(\mathcal{A})$ is sufficient. However $\mathcal{B}$ is not pairwise sufficient, because $\mathcal{B}$ does not separate $\{0\}$ and $\{1\}$ which are parts of the smallest sufficient statistic (Theorem 3.3). Note that $\mathcal{B}$ does not contain supports of P_0 and P_1.

Then next theorem characterizes PSS subfields by the existence of an estimator having variance which is smaller than or equal to a given unbiased estimator.

Theorem 4.3. Let $\mathcal{E}$ be weakly dominated and $\mathcal{B}$ be a subfield of $\mathcal{A}$. Then $\mathcal{B}$ is PSS if and only if for any real valued $t\in\cap\{L^1(\mathcal{X},\mathcal{A},P_\theta);\theta\in\Theta\}$ there exists a $\hat{\mathcal{B}}(\mathcal{A})$-measurable real valued function g such that

(1) $[g(x)<r]\cap S(\theta)\in\bar{\mathcal{B}}(\mathcal{A})$,

and

(2) $P_\theta([g(x)\neq E(t/\mathcal{B},P_\theta)(x)])=0$ for all $\theta\in\Theta$ and all real numbers r.

Here $L^1(\mathcal{X},\mathcal{A},P_\theta)$ is the totality of (equivalence classes of) functions which are integarble w.r.t. P_θ.

<u>Proof</u>. Let us assume that $\mathcal{B}$ is PSS. By Lemma 2.3 $\hat{\mathcal{B}}(\mathcal{A})$ is sufficient, so there exists a $\hat{\mathcal{B}}(\mathcal{A})$-measurable function g such that

$$\int_B t(x)\,dP_\theta=\int_B g(x)\,dP_\theta \text{ for all } \theta\in\Theta \text{ and } B\in\hat{\mathcal{B}}(\mathcal{A}).$$

For any $B\in\hat{\mathcal{B}}(\mathcal{A})$ and P_θ there exist sets $B(\theta)\in\mathcal{B}$ and $N(\theta)\in\mathcal{A}$ such that $B\Delta B(\theta)\subset N(\theta)$ and $P_\theta(N(\theta))=0$. Then we have

$$E(t/\mathcal{B},P_\theta)(x)=g(x) \text{ a.e. } P_\theta,$$

because

$$\int_B E(t/\mathcal{B},P_\theta)(x)\;dP_\theta = \int_{B(\theta)} E(t/\mathcal{B},P_\theta)(x)\;dP_\theta$$

$$=\int_{B(\theta)} t(x)\;dP_\theta = \int_B t(x)\;dP_\theta = \int_B g(x)\;dP_\theta.$$

Since $[g(x)<r]\in\hat{\mathcal{B}}(\mathcal{A})$ for any real number r and $P_\theta\in\mathcal{P}$, there exist sets $B'(\theta)\in\mathcal{B}$ and $N'(\theta)\in\mathcal{A}$ such that

$$[g(x)<r]\Delta B'(\theta)\subset N'(\theta),\; P_\theta(N'(\theta))=0. \tag{4.1}$$

Let m be an equivalent majorizing measure for $\mathcal{E}$ and $p(\cdot,\theta)$ be a version of the density of P_θ w.r.t. m such that $[p(x,\theta)>0]\in\mathcal{B}$.
Then by (4.1) it follows that

$$([p(x,\theta)>0]\cap[g(x)<r])\Delta([p(x,\theta)>0]\cap B'(\theta))$$
$$\subset[p(x,\theta)>0]\cap N'(\theta). \tag{4.2}$$

And

$$m([p(x,\theta)>0]\cap N'(\theta))=0 \tag{4.3}$$

holds because

$$0=P_\theta([p(x,\theta)>0]\cap N'(\theta))=\int_{[p(x,\theta)>0]\cap N'(\theta)} p(x,\theta)\ dm(x).$$

From (4.2) and (4.3) we have

$$[p(x,\theta)>0]\cap[g(x)<r]\in\bar{\mathcal{B}}(\mathcal{A}).$$

Therefore for any support $S(\theta)$ of P_θ it will be proved that

$$[h(x)<r]\cap S(\theta)\in\bar{\mathcal{B}}(\mathcal{A}).$$

Next we prove the "if" part. For $t=1$ we have a $\hat{\mathcal{B}}(\mathcal{A})$-measurable function g which satisfies the conditions of the "if" part. Hence $g(x)=1$ a.e. m because $m\sim\mathcal{P}$. So we can assume that, without loss of generality, $g(x)\leq 1$ for all $x\in\mathcal{X}$. Then it follows that

$$[dP_\theta/dm(x)>0]=[g(x)<2]\cap[dP_\theta/dm(x)>0]\in\bar{\mathcal{B}}(\mathcal{A}).$$

Hence there exist sets $B(\theta)\in\mathcal{B}$ and $N(\theta)\in\mathcal{A}$ such that

$$[dP_\theta/dm(x)>0]\,\Delta\, B(\theta)\subset N(\theta),\ N(\theta)\in N(\mathcal{P}).$$

Define a function $p(\cdot,\theta)$ by

$$p(x,\theta)=\begin{cases} dP_\theta/dm(x) & \text{if } x\in B(\theta)\cap[dP_\theta/dm(x)>0],\\ =1 & \text{if } x\in B(\theta)-[dP_\theta/dm(x)>0],\\ =0 & \text{otherwise.}\end{cases}$$

Then it can be checked that $[p(x,\theta)>0]=B(\theta)\in\mathcal{B}$ and

$$m([p(x,\theta)\neq dP_\theta/dm(x)])=0.$$

So $p(\cdot,\theta)$ is a version of the density of P_θ w.r.t. m. For any $t\in\cap\{L^1(\mathcal{X},\mathcal{A},P_\theta);\theta\in\Theta\}$, by the assumption, there exists a

$\hat{\mathcal{B}}(\mathcal{A})$-measurable function g such that, for all $\theta\in\Theta$,

$$P_\theta([g(x)\neq E(t/\mathcal{B},P_\theta)(x)])=0.$$

For any $B\in\hat{\mathcal{B}}(\mathcal{A})$ and $P_\theta\in\mathcal{P}$ there exist sets $B(\theta)\in\mathcal{B}$ and $N(\theta)\in\mathcal{A}$ such that $B\Delta B(\theta)\subset N(\theta)$, $P_\theta(N(\theta))=0$. Then we have

$$\int_B g(x)\ dP_\theta = \int_{B(\theta)} g(x)\ dP_\theta = \int_{B(\theta)} E(t/\mathcal{B},P_\theta)(x)\ dP_\theta$$

$$= \int_{B(\theta)} t(x)\ dP_\theta = \int_B t(x)\ dP_\theta = \int_B E(t/\hat{\mathcal{B}}(\mathcal{A}),P_\theta)\ dP_\theta.$$

Hence it folllows that $g(x) = E(t/\hat{\mathcal{B}}(\mathcal{A}),P_\theta)(x)$ a.e. P_θ. Therefore $\hat{\mathcal{B}}(\mathcal{A})$ is sufficient. Then by Lemma 4.1 $\mathcal{B}$ is pairwise sufficient.

Remark 4.2. Even if g is $\hat{\mathcal{B}}(\mathcal{A})$-measurable, without the assumption about supports for $\mathcal{B}$, it does not follow that $[h(x)<r]\cap[dP_\theta/dm(x)>0]\in\bar{\mathcal{B}}(\mathcal{A})$ for all $P_\theta\in\mathcal{P}$ and all real numbers r. In Remark 4.1 let $\mathcal{B}$ be the subfield generated by all singletons but 0. Then $\mathcal{B}$ is pairwise sufficient by Theorem 3.7 (2), and $\hat{\mathcal{B}}(\mathcal{A})=\mathcal{A}$. In the "only if" part of Theorem 4.3, we can take t as the indicator of $\{0\}$. Then the corresponding function g must coincide with t. However $[g(x)<2]\cap[dP_\theta/dm(x)>0]=\{0\}$ does not belong to $\bar{\mathcal{B}}(\mathcal{A})=\mathcal{B}$. Here m is the counting measure. As is easily seen the converse of the above statement holds true in any case: If $[g(x)<r]\cap[dP_\theta/dm(x)>0]\in\bar{\mathcal{B}}(\mathcal{A})$ for all $\theta\in\Theta$ and all real numbers r, then g is $\hat{\mathcal{B}}(\mathcal{A})$-measurable.

Next theorem gives a characterization of PSS in terms of a Rao-Blackwell theorem.

Theorem 4.4. Let $\mathcal{E}$ be weakly dominated and let $\mathcal{B}$ be a sub-field of $\mathcal{A}$. Then $\mathcal{B}$ is PSS if and only if for any real valued function $t\in\cap\{L^2(\mathcal{X},\mathcal{A},P_\theta);\theta\in\Theta\}$ there exists a $\hat{\mathcal{B}}(\mathcal{A})$-measurable real valued function g such that

(1) $[g(x)<r]\cap S(\theta)\in\bar{\mathcal{B}}(\mathcal{A})$,
(2) $\int g(x)\ dP_\theta = \int t(x)\ dP_\theta$, and

(3) $\int g^2(x)\ dP_\theta \le \int t^2(x)\ dP_\theta$

for all $\theta\in\Theta$ and all real numbers r.

<u>Proof</u>. "Only if" part. By Theorem 4.3 there exists a $\hat{\mathcal{B}}(\mathcal{A})$-measurable function g such that $[g(x)<r]\cap S(\theta)\in\bar{\mathcal{B}}(\mathcal{A})$ and $P_\theta([g(x)\neq E(t/\mathcal{B},P_\theta)(x)])=0$ for all $P_\theta\in\mathcal{P}$ and all real numbers r. Hence we have

$$\int g(x)\ dP_\theta = \int E(t/\mathcal{B},P_\theta)(x)\ dP_\theta = \int t(x)\ dP_\theta$$

for all $\theta\in\Theta$. By Jensen's inequality

$$|E(t/\mathcal{B},P_\theta)|^2 \le E(t^2/\mathcal{B},P_\theta) \text{ a.e. } P_\theta.$$

So we have

$$g^2(x) \le E(t^2/\mathcal{B},P_\theta)(x) \text{ a.e. } P_\theta.$$

Then it follows that

$$\int g^2(x)\ dP_\theta \le \int E(t^2/\mathcal{B},P_\theta)\ dP_\theta = \int t^2(x)\ dP_\theta \quad \text{for all } \theta\in\Theta.$$

"If" part. Since $\hat{\mathcal{B}}(\mathcal{A})$ has the Rao-Blackwell property, $\hat{\mathcal{B}}(\mathcal{A})$ is pairwise sufficient by Theorem 4.1. But $\hat{\mathcal{B}}(\mathcal{A})$ is actually sufficient by Lemma 2.3 and $(\hat{\hat{\mathcal{B}}}(\mathcal{A}))(\mathcal{A}) = \hat{\mathcal{B}}(\mathcal{A})$. On the other hand it follows that

$$S(\theta)=\{\cup\{[g(x)<r];\ r \text{ is rational}\}\}\cap S(\theta)=\cup\{[g(x,\theta)<r]\cap S(\theta);\ r \text{ is rational}\}\in\bar{\mathcal{B}}(\mathcal{A}).$$

Then there exists $T(\theta)\in\mathcal{B}$ such that $T(\theta)\Delta\ S(\theta)\in N(\mathcal{P})$, and hence $T(\theta)$ is a support of P_θ belonging to $\mathcal{B}$. By Lemma 4.1 $\mathcal{B}$ is pairwise sufficient.

<u>Remark 4.3</u>. In the proof of the above Theorem we used Lemma 2.3. But Lemma 2.3 does not hold without the assumption of weak domination, and the "only if" part of Theorem 4.4 is not correct in a general case. Let us take up the following example. Let $\mathcal{X}=(-1,1)$, and $\mathcal{A}$ be the Borel field of subsets of $\mathcal{X}$ and $\Theta=[0,1)$. For each $\theta\in\Theta$ let $p(x,\theta)$ satisfy the following

conditions;

(1) $p(\theta,\theta)>0$, $p(-\theta,\theta)>0$ for all $\theta\in\Theta$,
(2) $p(x,\theta)=0$ for $x\neq\theta$ and $-\theta$,
(3) $p(\theta,\theta)+p(-\theta,\theta) = 1$ for all $\theta\in\Theta-\{0\}$ and $p(0,0)=1$.

We define P_θ, for each $\theta\in\Theta-\{0\}$, as a two point probability measure on $(\mathcal{X},\mathcal{A})$ with probabilities $p(\theta,\theta)$ on θ and $p(-\theta,\theta)$ on $-\theta$. P_0 is the point probability measure on 0. Define $\mathcal{P}=\{P_\theta;\theta\in\Theta\}$ and $\mathcal{B}=\{A\in\mathcal{A};\ A$ is countable and $A=-A\}\cup\{A\in\mathcal{A};\mathcal{X}-A$ is countable and $A=-A\}$, $\mathcal{C}=\{A\in\mathcal{A};A=-A\}$, where $-A=\{-x;x\in A\}$. Then $\mathcal{B}$ is PSS. And $\mathcal{C}$ is sufficient if and only if $p(\theta,\theta)$ is a Borel measurable function on Θ. Note that $\hat{\mathcal{B}}(\mathcal{A})=\mathcal{C}$. Hence, if we take $p(\theta,\theta)$ as a non-Borel measurable function, $\hat{\mathcal{B}}(\mathcal{A})$ is not sufficient. If the "only if" part of Theorem 4.4 holds true for this $\mathcal{B}$, we shall have the following relation using only "unbiasedness" condition $p(\theta,\theta)=(g(\theta)-t(-\theta))/(t(\theta)-t(-\theta))$ for all $\theta\in\Theta$. But this is a contradiction if $p(\theta,\theta)$ is not a Borel measurable function.

Next we will introduce another kind of representation of "sufficiency" in terms of tests.

Definition 4.2. Let $\mathcal{E} = (\mathcal{X},\mathcal{A},\mathcal{P}=\{P_\theta;\theta\in\Theta\})$ be any experiment. Let $\mathcal{B}$ be a subfield of $\mathcal{A}$. Then $\mathcal{B}$ is said to be *test-sufficient* if for any $\mathcal{A}$-measurable test function f, that is a $\mathcal{A}$-measurable function whose range is contained in $[0,1]$, there exists a $\mathcal{B}$-measurable test function g such that

$$\int f(x)\ dP_\theta = \int g(x)\ dP_\theta \text{ for all } \theta\in\Theta.$$

Theorem 4.5. Let $\mathcal{E}$ be a majorized experiment and $\mathcal{B}$ be test-sufficient. Then $\mathcal{B}$ is PSS.

Proof. We show first $\mathcal{B}$ is pairwise sufficient. Take any $P_1=P_{\theta_1}$ and $P_2=P_{\theta_2}$, and let's define $m=P_1+P_2$. Suppose that t is the indicator function of $[p_1(x)<cp_2(x)]$, where c is a positive constant and p_i is a non-negative version of the density of P_i w.r.t. m, $i=1,2$. As $\mathcal{B}$ is test-sufficient there exists a $\mathcal{B}$-measurable test function s such that

$$\int s(x)\ dP_i = \int t(x)\ dP_i,\ i=1,2.$$

Let's define $g=t-s$, $h= g(cp_2-p_1)$. Then we have

$$[g(x)>0]\subset[cp_2(x)-p_1(x)>0],$$
$$[g(x)<0]\subset[cp_1(x)-p_2(x)\leq 0].$$

So $h\geq 0$. On the other hand

$$\int h(x)\ dm = \int g(cp_2-p_1)\ dm = c\int g\ dP_2 - \int g\ dP_1 = 0.$$

This shows $h(x)=0$ a.e. m. Since $[g(x)>0]\subset[h(x)>0]$, $g(x)\leq 0$ a.e. m. Thus we have $t=s$ a.e. m. Let $\bar{\mathcal{B}}(m)$ be the completion of $\mathcal{B}$ w.r.t. m, i.e., $\bar{\mathcal{B}}(m)=\{A\in\mathcal{A};$ there exists $B\in\mathcal{B}$ such that $m(A\Delta B)=0\}$. Then it follows that

$$\bar{f}_1 = (p_1/p_2)I_{S_2} + I_{S_2{}^c},\ \bar{f}_2 = I_{S_2},\ p= p_2+p_1 I_{S_2{}^c} \text{ are}$$

$\bar{\mathcal{B}}$-measurable, where $S_2=[p_2(x)>0] = \bigcup_{n=1}^{\infty}[p_1(x)<p_2(x)]$. Hence there exist $\mathcal{B}$-measurable functions f_1, f_2, q such that $\bar{f}_i=f_i$ a.e. m and $p=q$ a.e. m. It is easily checked that $p\bar{f}_i=p_i$ a.e. m., $i=1,2$. So qf_i is a $\mathcal{B}$-measurable density of P_i w.r.t. m, $i=1,2$. This implies $\mathcal{B}$ is pairwise sufficient.

Next we prove $\mathcal{B}$ contains supports. Take any $\theta\in\Theta$ and let m be an equivalent dominating measure for $\mathcal{P}$. Let's define $t(x)=I_{T_\theta}(x)$, where $T_\theta=[dP_\theta/dm(x)>0]$. Then, by test sufficiency of $\mathcal{B}$, there exists a $\mathcal{B}$-measurable test function s such that

$$\int s(x)\ dQ = \int t(x)\ dQ$$

for all $Q\in\mathcal{P}$. So $s(x)=1$ a.e. P_θ on T_θ. Then, by

$$0=\int t(1-s)\ dQ = \int t(1-s)\ dP_\theta/dm\ dm,$$

we have $t=ts$ a.e. m. For any $Q\in\mathcal{P}$ it follows that

$$\int ts\ dQ = \int s\ dQ = \int s\ dQ,$$

so $s=ts$ a.e. Q. Thus $t=s$ a.e. Q for all $Q\in\mathcal{P}$. Let's denote $B_\theta=[s(x)=1]$. Then $T_\theta\Delta B_\theta\subset[t\neq s]$, which implies $T_\theta\Delta B_\theta\in N(\mathcal{P})=N(m)$.

Then $f(x,\theta)$ which is defined by

$$f(x,\theta) = dP_\theta/dm(x) \quad \text{if } x\in T_\theta\cap B_\theta,$$
$$= 1 \quad \text{if } x\in B_\theta - T_\theta,$$
$$=0 \quad \text{otherwise},$$

satisfies $[f(x,\theta)>0] = B_\theta\in\mathcal{B}$ and $[f(x,\theta)\neq dP_\theta/dm(x)] = T_\theta\Delta B_\theta$. So $f(\cdot,\theta)$ is a version of the density of P_θ w.r.t. m. Hence B_θ is a support of P_θ which belongs to $\mathcal{B}$.

The following two colloralies follow from above results and the proof is omitted. They explain the difference between pairwise sufficiency and PSS, and show their relations to other notions. At the same time we should here recall that PSS is characterized by a Neyman factorization but pairwise sufficiency is not. In the following, for a subfield $\mathcal{B}$, the strong self-completion $\bar{\mathcal{B}}(\mathcal{B})$ of $\mathcal{B}$ is defined by

$$\{A\in\mathcal{A};\text{there exist } B\in\mathcal{B} \text{ and } N\in\mathcal{B} \text{ such that } A\Delta B\subset N\ [\mathcal{P}] \text{ and } P_\theta(N)=0 \text{ for all } \theta\in\Theta\}.$$

Note that $\hat{\mathcal{B}}(\mathcal{A})=\bar{\mathcal{B}}(\mathcal{A})$ if $\mathcal{B}$ is pairwise sufficient. This is similar to Lemma 2.4.

Corollary 4.2. Let $\mathcal{E}$ be weakly dominated and let $\mathcal{B}$ be a subfield. Then the following conditions are all equivalent.

(a) $\mathcal{B}$ is PSS.

(b) $\tilde{\mathcal{B}}(\mathcal{B})$ is PSS.

(c) $\tilde{\mathcal{B}}(\mathcal{B})$ is sufficient.

(d) $\tilde{\mathcal{B}}(\mathcal{B})$ is test-sufficient.

(e) $\tilde{\mathcal{B}}(\mathcal{B})$ has the Rao-Blackwell property and contains supports.

(f) For any real valued $f\in\cap\{L^1(\mathcal{X},\mathcal{A},P_\theta);\theta\in\Theta\}$ there exists a $\tilde{\mathcal{B}}(\mathcal{B})$-measurable function g such that

$$P_\theta([g(x)\neq E(f/\mathcal{B},P_\theta)(x)])=0$$

for all $\theta\in\Theta$ and there exists a support $S(\theta)$ of P_θ such that $S(\theta)\cap[h(x)<r]\in\bar{\mathcal{B}}(\mathcal{B})$ for all real numbers r.

(g) $\tilde{\mathcal{B}}(\mathcal{B})$ has the Rao-Blackwell property in which we can take h as an existing $\tilde{\mathcal{B}}(\mathcal{B})$-measurable real valued function satisfying the following: For any P_θ there exists a version $p(\cdot,\theta)$ of the density of P_θ w.r.t. m such that $[h(x)<r]\cap[p(x,\theta)>0]\in\bar{\mathcal{B}}(\mathcal{B})$ for all real numbers r.

Corollary 4.3. Let $\mathcal{E}$ be weakly dominated and let $\mathcal{B}$ be a subfield. Then the following conditions are all equivalent.

(a) $\mathcal{B}$ is pairwise sufficient.

(b) $\tilde{\mathcal{B}}(\mathcal{A})$ is PSS.

(c) $\tilde{\mathcal{B}}(\mathcal{A})$ is sufficient.

(d) $\tilde{\mathcal{B}}(\mathcal{A})$ is test-sufficient.

(e) $\tilde{\mathcal{B}}(\mathcal{A})$ is pairwise sufficient.

(f) $\tilde{\mathcal{B}}(\mathcal{A})$ has the Rao-Blackwell property.

4.2. Converse of a theorem of Lehmann and Scheffé

It is important to know whether the Rao-Blackwellization actually provides a UMVU estimator. To consider this problem let us recall the following definition of completeness: a subfield $\mathcal{B}$ is said to be quadratically *complete* if $u=0$ a.e. P_θ for all $\theta\in\Theta$ whenever u is $\mathcal{B}$-measurable, and

$$\int u(x)\ dP_\theta=0\ ,\ \int u^2(x)\ dP_\theta<\infty \text{ for all } \theta\in\Theta.$$

A real valued function g of the parameter is said to be quadratically, unbiasedly *estimable* if there exists an $\mathcal{A}$-measurable real valued estimator u such that

$$\int u(x)\ dP_\theta= g(\theta) \text{ and } \int u^2(x)\ dP_\theta<\infty \text{ for all } \theta\in\Theta.$$

The following theorem is a well known theorem which is due to Lehmann and Scheffé(1950). By the assumption of completeness the estimator obtained through the Rao-Blackwellization of an unbiased estimator is a UMVU estimator as is shown in the proof.

Theorem 4.6. Let $\mathcal{E}=(\mathcal{X},\mathcal{A},\mathcal{P}=\{P_\theta;\theta\in\Theta\})$ be any experiment and let $\mathcal{B}$ be quadratically complete and sufficient. Then for any quadratically, unbiasedly estimable g there exists a $\mathcal{B}$-

measurable estimator which is a UMVU estimator of g.

Proof. Let t be an estimator such that $\int t(x)\, dP_\theta = g(\theta)$ and $\int t^2(x)\, dP_\theta < \infty$ for all $\theta \in \Theta$. Let us define $h(x) = E(t/\mathcal{B})(x)$. Then it holds that, for all $\theta \in \Theta$, $\int h(x)\, dP_\theta = \int t(x)\, dP_\theta = g(\theta)$ and $\int h^2(x)\, dP_\theta \leq \int t^2(x)\, dP_\theta$ by Jensen's inequality.

Take any unbiased estimator s of g such that $\int s^2(x)\, dP_\theta < \infty$ for all $\theta \in \Theta$. Then it follows that $\int h(x)\, dP_\theta = \int s(x)\, dP_\theta = \int E(s/\mathcal{B})(x)\, dP_\theta$ for all $\theta \in \Theta$. By the assumption of completeness of $\mathcal{B}$ we have $h(x) = E(s/\mathcal{B})(x)$ a.e. P_θ for all $\theta \in \Theta$. This implies, for all $\theta \in \Theta$, $\int h^2(x)\, dP_\theta \leq \int s^2(x)\, dP_\theta$ again by Jensen's inequality. Hence h is a $\mathcal{B}$-measurable UMVU estimator of g.

It is the aim of this section to show conversely if any quadratically, unbiasedly estimable parameter function has a UMVU estimator then there exists a quadratically complete sufficient subfield. This implies the existence of a complete sufficient subfield is equivalent to the existence of a UMVU estimator for each quadratically, unbiasedly estimable parameter function. Hence in cases where there are no complete sufficient subfields, for some parameter function we can not get any UMVU estimator.

Bahadur (1957) proved this converse under the assumption of domination. We extend this to the case of majorized experiments. This is due to Torgersen (1988).

Let us introduce a few notations. Let $\mathcal{E} = (\mathcal{X}, \mathcal{A}, \mathcal{P} = \{P_\theta; \theta \in \Theta\})$ be an experiment. Then

$U = \cap\{L^2(\mathcal{X}, \mathcal{A}, P_\theta);\theta \in \Theta\}$,

$T = \{t \in U; \int t^2(x)\, dP_\theta \leq \int u^2(x)\, dP_\theta$ for all $\theta \in \Theta$ whenever $u \in U$ and $\int t(x)\, dP_\theta = \int u(x)\, dP_\theta$ for all $\theta \in \Theta\}$,

$\tilde{T} = \{t \in T; P_\theta(|t| \leq g(\theta)) = 1$ for all θ and for some real valued function g on $\Theta\}$,

$N = \{u; \int u(x)\, dP_\theta = 0$ for all $\theta \in \Theta\}$, and

$\mathcal{B} = \{B \in \mathcal{A}; I_B \in T\}$.

Next Lemma shows that an estimator is a UMVU estimator if and only if it is orthogonal to all unbiased estimators of 0.

Lemma 4.2. For any $t \in U$, t belongs to T if and only if $tz \in N$ whenever $z \in N \cap U$.

Proof. "If" part. Take any $u \in U$ such that $\int u(x)\, dP_\theta = \int t(x)\, dP_\theta$ for all $\theta \in \Theta$. Then $u - t \in N \cap U$. By assumption

$$0=\int t(x)(u(x) - t(x))\, dP_\theta = \int t(x)u(x)\, dP_\theta - \int t^2(x)\, dP_\theta$$

$$\leq \left(\int t^2(x)\, dP_\theta\right)^{1/2}\left(\int u^2(x)\, dP_\theta\right)^{1/2} - \int t^2(x)\, dP_\theta$$

for all $\theta \in \Theta$, which implies $\int u^2(x)\, dP_\theta \geq \int t^2(x)\, dP_\theta$ for all $\theta \in \Theta$.

"Only if" part. Take any $z \in N \cap U$. Then for any real number r,

$$\int (t(x) + rz(x))\, dP_\theta = \int t(x)\, dP_\theta \text{ for all } \theta \in \Theta.$$

Since $t \in T$ we have

$$\int (t(x) + rz(x))^2\, dP_\theta \geq \int t^2(x)\, dP_\theta, \text{ that is,}$$
$$r^2 \int z^2(x)\, dP_\theta + r\int z(x)t(x)\, dP_\theta \geq 0$$

for all $r \in R^1$ and $\theta \in \Theta$. If $\int z^2(x)\, dP_\theta = 0$ then $z(x)=0$ a.e. P_θ, and so $\int z(x)t(x)\, dP_\theta = 0$. If $\int z^2(x)\, dP_\theta > 0$ then, by the discriminant criterion, it follows that $0 \leq \left(\int t(x)z(x)\, dP_\theta\right)^2 \leq 0$. Hence $\int t(x)z(x)\, dP_\theta = 0$.

Theorem 4.7. (a) $U, T, \tilde{T}$ and N are linear spaces. If $t_1 \in \tilde{T}$ and $t_2 \in T$ then $t_1 t_2 \in T$. If $t_1, t_2 \in T$ and $\int t_1\, dP_\theta = \int t_2\, dP_\theta$ for all $\theta \in \Theta$ then $t_1 = t_2$ a.e. P_θ for all $\theta \in \Theta$.
(b) $\mathcal{B}$ is a subfield. Any $\mathcal{B}$-measurable estimator which belongs to U is a UMVU estimator. $\tilde{T}$ is the totality of bounded $\mathcal{B}$-measurable functions.
(c) $\mathcal{E}$ admits a quadratically complete sufficient subfield if and only if $P_\theta(A)$, for each $A \in \mathcal{A}$, possesses a non-negative UMVU estimator.

Proof. (a) These follow from Lemma 4.2.

(b) By (a) it is shown that $\mathcal{B}$ is a σ-algebra. Suppose that $u \in U$ is non-negative and $\mathcal{B}$-measurable. Then there exists a monotone increasing sequence $\{t_n\}$ of non-negative $\mathcal{B}$-measurable simple functions which converges to u. So we have

$$\int z^+u\, dP_\theta = \lim_{n\to\infty} \int z^+t_n\, dP_\theta = \lim_{n\to\infty} \int z^-t_n\, dP_\theta = \int z^-u\, dP_\theta,$$

and hence $\int zu\, dP_\theta = 0$ for all $\theta\in\Theta$ and $u\in N\frown U$. This shows that $u\in T$. Here we used the notations $z^+(x)=max\{z(x),0\}$ and $z^-(x)= max\{-z(x),0\}$.

Conversely suppose that $t\in\tilde{T}$. Then, by (a), it follows that $1, t, t^2, \ldots$ all belong to $\tilde{T}$. By the Weierstrass approximation theorem any continuous function of t belongs to $\tilde{T}$. In particular $min\{n(t-c)^+, 1\}\in\tilde{T}$ for any integer n and any constant c, which implies $I_{[t>c]}\in T$ by letting $n\to\infty$. By the definition of $\mathcal{B}$ $[t(x)>c]\in\mathcal{B}$, and hence t is $\mathcal{B}$-measurable.

(c) Suppose that, for each $A\in\mathcal{A}$, t_A is a UMVU estimator of $\theta\to P_\theta(A)$. Because $t_A + t_{\mathcal{X}-A}$ is a UMVU estimator of 1, we have $t_A + t_{\mathcal{X}-A} = 1$ a.e. P_θ for all $\theta\in\Theta$. This shows $t_A\in\tilde{T}$. By (b) t_A is $\mathcal{B}$-measurable. For any $B\in\mathcal{B}$ it follows that, since $t_A - I_A\in N\frown U$,

$$\int_B t_A\, dP_\theta = \int I_B(t_A - I_A)\, dP_\theta + \int I_A I_B\, dP_\theta = P_\theta(A\frown B),$$

for all $\theta\in\Theta$. This implies that t_A is a conditional probability of A given $\mathcal{B}$ which is common to all θ. Hence $\mathcal{B}$ is sufficient for $\mathcal{E}$. If $t\in N\frown U$ and is $\mathcal{B}$-measurable then t is a UMVU estimator of 0 by (b). So we have $t=0$ a.e. P_θ for all θ. Hence $\mathcal{B}$ is quadratically complete.

The converse part is easy to prove, so it is omitted.

Now we introduce the notion of minimality of the density. Take any $\theta_0\in\Theta$ and a finite subset F of Θ such that $\theta_0\in F$. A version u of the density of P_{θ_0} w.r.t. $P_F \equiv\sum\{P_\theta;\theta\in F\}$ is said to be *minimal* if $u(x)\geq 0$ for all $x\in\mathcal{X}$ and for any other non-negative version v of the density it holds $v(x)\geq u(x)$ a.e. P_θ for all $\theta\in\Theta$. The existence of such a minimal version of the density is guaranteed by the condition that $\mathcal{E}$ is majorized. Let m be a majorizing measure of $\mathcal{E}$. Let $g(\cdot,\theta)$ be the density of P_θ w.r.t. m. Suppose that $g(x,\theta)\geq 0$ for all $x\in\mathcal{X}$ and $\theta\in\Theta$. Define

$$g(x,F)=\sum\{g(x,\theta);\theta\in F\},$$

where F is a finite subset of Θ. Take any $\theta_0\in F$ and define

$$u(x;\theta_0,F) = g(x,\theta_0)/g(x,F) \qquad \text{if } g(x,F)>0,$$
$$= 0 \qquad \text{if } g(x,F)=0.$$

Then $u(x) \equiv u(x;\theta_0,F)$ is a minimal version of the density of P_{θ_0} w.r.t. P_F. To show this suppose that v is any other non-negative version of the density of P_{θ_0} w.r.t. P_F. Then, for any θ, it follows that

$$P_\theta(v(x)<u(x)) = \int_{[v(x)<u(x)]} g(x,\theta)\ dm$$

$$= \int_{[v(x)<u(x)]} \{g(x,\theta)/g(x,F)\}\ g(x,F)\ dm$$

$$= \int_{[v(x)<u(x)]} g(x,\theta)/g(x,F)\ dP_F = 0,$$

since $P_F([v(x) \neq u(x)])=0$.

Lemma 4.3. Suppose that u is a minimal version of the density of P_{θ_0} w.r.t. $P_F = \sum\{P_\theta;\theta\in F\}$, where F is a finite subset of Θ containing θ_0. If the parameter function $\theta \to \int u(x)\ dP_\theta$ has a UMVU estimator then $u \in T$.

Proof. Let $t \in T$ be such that $\int t(x)\ dP_\theta = \int u(x)\ dP_\theta$ for all $\theta\in\Theta$. Then $t(x) = u(x)$ a.e. P_F because

$$\int (t(x) - u(x))^2\ dP_F = \int t^2(x)\ dP_F + \int u^2(x)\ dP_F$$

$$- 2\int t(x)u(x)\ dP_F \leq 2\int u^2(x)\ dP_F - 2\int t(x)u(x)\ dP_F$$

$$= 2\int (u(x) - t(x))u(x)\ dP_F = 2\int (u(x) - t(x))\ dP_{\theta_0} = 0.$$

Hence t and $|t|$ are versions of the density of P_{θ_0} w.r.t. P_F. By the assumption of minimality of u, $u(x) \leq |t(x)|$ a.e. P_θ, and hence $u^2(x) \leq t^2(x)$ a.e. P_θ for all $\theta\in\Theta$.

On the other hand $\int t^2(x)\ dP_\theta \leq \int u^2(x)\ dP_\theta$ for all $\theta\in\Theta$, and hence $\int t^2(x)\ dP_\theta = \int u^2(x)\ dP_\theta$ for all $\theta\in\Theta$. Thus $u \in T$.

Theorem 4.8. Let $\mathcal{E} = (\mathcal{X}, \mathcal{A}, \mathcal{P} = \{P_\theta;\theta\in\Theta\})$ be a majorized experiment. Then $\mathcal{E}$ has a quadratically complete sufficient subfield if and only if for every quadratically, unbiasedly estimable parameter function there exists a UMVU estimator of it.

Proof. Only "if" part is to be proved. By Lemma 4.3 $u(x;\theta,F)\in T$ if F is a finite subset of Θ such that $\theta\in F$. In particular $I_{[g(x,\theta)>0]}(x) = u(x;\theta,\{\theta\})$ belongs to T. By Theorem 4.7 (b) $[g(x,\theta)>0]\in\mathcal{B}$ for each $\theta\in\Theta$. Take any UMVU estimator t_A of $\theta\to P_\theta(A)$ whose existence is guaranteed by the assumption. If we can show that $t_A(x)\geq 0$ for all $x\in\mathcal{X}$, then by Theorem 4.7 (c) the proof is completed. Take any $\theta_0\in\Theta$ and define $B_0=[g(x,\theta_0)>0]$. Take any version $E(I_A/\mathcal{B},P_{\theta_0})(x)$ of the conditional probability whose values lie between 0 and 1. By Theorem 4.7 (a) and (b), $v(x)\equiv I_{B_0}(x)t_A(x)$ and $w(x)\equiv I_{B_0}(x)E(I_A/\mathcal{B},P_{\theta_0})(x)$ belong to T. Take any $\theta\in\Theta$ and denote $F=\{\theta,\theta_0\}$. Then $g(x,\theta)/g(x,\theta_0) = u(x;\theta,F)/u(x;\theta_0,F)$ on B_0. Then it follows that

$$\int v(x)\ dP_\theta = \int_{B_0} t_A(x)\ dP_\theta$$

$$=\int_{B_0} I_A(x)\ dP_\theta \quad \text{(because } I_{B_0}(t_A-I_A)\in N)$$

$$=\int_{B_0} I_A(x)g(x,\theta)\ dm = \int_{B_0} I_A(g(x,\theta)/g(x,\theta_0))g(x,\theta_0)\ dm$$

$$=\int_{B_0} E(I_A/\mathcal{B},P_{\theta_0})(x)(g(x,\theta)/g(x,\theta_0))g(x,\theta_0)\ dm \quad \text{(because}$$

$u(x;\theta,F)/u(x;\theta_0,F)$ is $\mathcal{B}$-measurable and $g(x,\theta_0)dm= dP_{\theta_0}$)

$$=\int_{B_0} E(I_A/\mathcal{B},P_{\theta_0})(x)\ dP_\theta = \int w(x)\ dP_\theta.$$

By Theorem 4.7 (a), $v(x) = w(x)$ a.e. $\mathcal{P}$ and hence $t_A(x) = E(I_A/\mathcal{B},P_{\theta_0})(x)$ a.e. P_{θ_0}. Since θ_0 is arbitrary we have $t_A(x)\geq 0$ a.e. $\mathcal{P}$. This completes the proof.

Corollary 4.4. Suppose that $\mathcal{E}$ is a majorized experiment. Then $\mathcal{E}$ admits a quadratically complete sufficient subfield if and only if $\theta\to P_\theta(A)$ possesses a UMVU estimator for each $A\in\mathcal{A}$.

4.3. Bibliographical notes

The original papers of the Rao-Blackwell theorem are Rao

(1945) and Blackwell(1947). Lehmann and Scheffé (1950) proved that the Rao-Blackwellization w.r.t. a complete sufficient statistic actually provides a UMVU estimator. Bahadur(1957, 1979), on the other hand, solved the converse problem in dominated case. Torgersen(1988) extended this to majorized experiments. This is the content of Section 4.2. Section 4.1 is mainly due to Yamada(1980) which considered the Rao-Blackwell property through PSS subfields.

Morimoto(1972) showed, in discrete cases, any test-sufficient subfield $\mathcal{B}$ induces the sufficient statistic $\mathcal{T}(\mathcal{B})$, which implies $\mathcal{B}$ is pairwise sufficient. Brown (1975) extended this and proved, under the same condition, that $\mathcal{B}$ includes a sufficient subfield. But the proof of Brown(1975) seems to have some difficulty.

Theorem 4.5 is important. It asserts that decision theoretic sufficiency restricted to testing problems implies PSS. We note that, without any restriction, test-sufficiency implies pairwise sufficiency as was shown in the first half of the proof of Theorem 4.5. The proof is due to Pfanzagl(1974). We also refer to Kudo(1957) and Strasser(1985). As was noted by Pfanzagl(1974), from this result it can be shown that any exhaustive subfield is pairwise sufficient (Bahadur(1955), Sacksteder(1967),and Heyer(1982)). Here a subfield $\mathcal{B}$ is said to be *exhaustive* if there exists a Markov kernel $M:(\mathcal{X},\mathcal{B})\longrightarrow(\mathcal{X},\mathcal{A})$ such that

$$\int M(x,A)\ dP_\theta/\mathcal{B} = P_\theta(A) \text{ for all } \theta\in\Theta \text{ and } A\in\mathcal{A}.$$

This definition is equivalent to *D-sufficiency* which was introduced by Roy and Ramamoorthi(1979). They showed any D-sufficient subfield includes a sufficient subfield. This is more general than the theorem of Bahadur(1955). The result of Brown(1975) in the preceding paragraph asserts, though restricted to discrete cases, more strongly mere test-sufficient subfield contains a sufficient subfield. The latter half of the proof of Theorem 4.5 is due to Mussmann(1971). Kusama and Fujii(1987) proved that in weakly dominated , but not dominated cases the smallest PSS subfield is not test-sufficient. This implies the converse of Theorem 4.5 does not hold.

Chapter 5. Common conditional probability for PSS and its applications

We have observed in previous chapters that PSS subfields have such properties as sufficient subfields have in dominated case. For example (a) there exists the smallest PSS, (b) A subfield is PSS if and only if it has a Neyman factorization, and (c) an improvement of a given unbiased estimator can be done by an unbiased estimator which is approximately measurable w.r.t. a PSS subfield. These properties were proved mainly using the notion of a pivotal measure.

The above observation suggests that the experiment induced by a PSS subfield is as "informative" as the original experiment. In this chapter in the first place we consider somewhat abstract problem of comparing these two experiments. Based on this examination we next construct an extended version of the conditional probability for PSS subfield which is common to all the parameters. Finally we apply this concept to unbiased estimation theory. It is no wonder that pivotal measure again plays important roles. For this it is enough to recall the following result in dominated experiment.

Theorem 5.1. Let $\mathcal{E}=(\mathcal{X},\mathcal{A},\mathcal{P}=\{P_\theta;\theta\in\Theta\})$ be dominated and let $\mathcal{B}$ be sufficient. Then it holds that

$$E(I_A/\mathcal{B})(x)=E(I_A/\mathcal{B},n)(x) \text{ a.e. } P_\theta$$

for all $\theta\in\Theta$, where $A\in\mathcal{A}$ and n is the pivotal measure defined by $n(A)=\sum_{i=1}^{\infty} c_i P_{\theta_i}(A)$ (Section 2.4).

Proof. This follows, since

$$n(A\cap B)=\sum c_i P_{\theta_i}(A\cap B)=\sum c_i \int_B E(I_A/\mathcal{B})\,dP_{\theta_i}=\int_B E(I_A/\mathcal{B})\,dn,$$

for all $A\in\mathcal{A}$, $B\in\mathcal{B}$, and n is equivalent to $\mathcal{P}$.

Main references for this chapter are Le Cam(1964), Torgersen(1979), and Yamada (1985,1988,1991).

5.1. Deficiency

We collect here fundamental notions from the theory of comparison of experiments used in this chapter.

Definition 5.1. Let $\mathcal{E}=(\mathcal{X},\mathcal{A},\mathcal{P}=\{P_\theta;\theta\in\Theta\})$ be any experiment.
(1) Let $\mathfrak{M}(\mathcal{A})$ be the space of all bounded signed measures on $(\mathcal{X},\mathcal{A})$. Then an ideal B of $\mathfrak{M}(\mathcal{A})$ is called a *band* if any subset C of B which has an upper bound in $\mathfrak{M}(\mathcal{A})$ has a least upper bound in B. Here an *ideal* B is a sublattice of $\mathfrak{M}(\mathcal{A})$ such that $\mu\in B$ if $|\mu|\leq|\nu|$ and $\nu\in B$. Here the order between signed measures is defined pointwise.
(2) *L-space* of $\mathcal{E}$, denoted by $L(\mathcal{E})$, is defined as the smallest band including $\mathcal{P}$ in $\mathfrak{M}(\mathcal{A})$. Consider the total variation metric on $L(\mathcal{E})$. Then the dual space of $L(\mathcal{E})$, which is denoted by $M(\mathcal{E})$, is called *M-space* of $\mathcal{E}$.
(3) Let $\mathcal{B}$ be a subfield of $\mathcal{A}$. Then a *transition* from $\mathcal{E}(\mathcal{B})$ to $\mathcal{E}$ is defined as a positive, linear and positively isometric mapping T on $L(\mathcal{E}(\mathcal{B}))$ into $L(\mathcal{E})$. Here positivity means $T(\mu)\geq 0$ whenever $\mu\geq 0$, and T is positively isometric if $\|T(\mu)\| = \|\mu\|$ if $\mu\geq 0$. The notation $\|\cdot\|$ denotes the total variation norm.
(4) The *deficiency*, which is denoted by $\delta(\mathcal{E}(\mathcal{B}),\mathcal{E})$, of $\mathcal{E}(\mathcal{B})$ relative to $\mathcal{E}$ is defined by

$$inf\{sup\{\|TP_\theta/\mathcal{B} - P_\theta\|;\theta\in\Theta\};T\},$$

where the infimum is taken over all the transitions from $\mathcal{E}(\mathcal{B})$ to $\mathcal{E}$.

The following theorem gives the concrete representation of the L-space of a majorized experiment which is similar to the famous theorem of Kakutani(1941).

Theorem 5.2. Let $\mathcal{E}$ be a majorized experiment and n be an equivalent majorizing measure. Then we have

$$L(\mathcal{E}) =\{f\cdot n;f\in L^1(\mathcal{X},\mathcal{A},n)\}.$$

Here $f\cdot n$ is the signed measure which has f as the Radon-Nikodym derivative w.r.t. n.

Proof. Let us denote

$$L(n)=\{f\cdot n; f\in L^1(\mathcal{X},\mathcal{A},n)\}.$$

Then $L(n)$ is a band in $\mathfrak{M}(\mathcal{A})$ and includes $\mathcal{P}$ (Schaefer(1974),p. 113,§8.3(ii)). Therefore $L(n)$ is the direct sum of $L(\mathcal{E})$ and a band L in $L(n)$ because $L(\mathcal{E})\subset L(n)$ (Schaefer(1974).p.62. Theorem 2.10). Hence any $f\cdot n\geqq 0$ in L is orthogonal to $P_\theta\cdot n$ for all $\theta\in\Theta$. Then $N=[f(x)>0]$ is a $\mathcal{P}$-null set, which implies $n(N)=0$ because $\mathcal{P}\sim n$. Hence $f\cdot n=0$ and $L=\{0\}$. This shows $L(\mathcal{E})=L(n)$.

Next Lemma shows $L(\mathcal{E})$ is the union of the L-spaces of countable subfamilies of $\mathcal{P}$.

<u>Lemma 5.1</u>. Let $\mathcal{E}$ be any experiment. Then for any $\mu\in\mathfrak{M}(\mathcal{A})$, μ belongs to $L(\mathcal{E})$ if and only if there exists a countable subfamily $\mathcal{P}_0$ of $\mathcal{P}$ such that μ belongs to the band in $\mathfrak{M}(\mathcal{A})$ generated by $\mathcal{P}_0$.

<u>Proof</u>. "Only if" part. Let π denote the totality of countable subfamilies of $\mathcal{P}$. We will show that

$$L(\mathcal{E})=\cup\{L(K); K\in\pi\},$$

where $L(K)$ denotes the band generated by K in $\mathfrak{M}(\mathcal{A})$. That $\cup\{L(K); K\in\pi\}$ is an ideal in $\mathfrak{M}(\mathcal{A})$ and is closed w.r.t. the total variation metric are easy to show. Hence $\cup\{L(K); K\in\pi\}$ is a band (Schaefer (1974),Chap.2,8.3). Clearly $\cup\{L(K); K\in\pi\}$ includes $\mathcal{P}$, and hence $\cup\{L(K); K\in\pi\}\supset L(\mathcal{E})$. The converse inclusion is obvious because $K\subset\mathcal{P}$ for all $K\in\pi$.

The "if" part is trivial.

<u>Corollary 5.1</u>. Let $\mathcal{E}=(\mathcal{X},\mathcal{A},\mathcal{P}=\{P_\theta;\theta\in\Theta\})$ be any experiment. Then for any $\mu\in\mathfrak{M}(\mathcal{A})$, μ belongs to $L(\mathcal{E})$ if and only if there exists a countable subfamily $\mathcal{P}_0$ of $\mathcal{P}$ such that $\mu\ll\mathcal{P}_0$.

This follows from Theorem 5.2 and Lemma 5.1.

<u>Lemma 5.2</u>. Let $\mathcal{E}=(\mathcal{X},\mathcal{A},\mathcal{P}=\{P_\theta;\theta\in\Theta\})$ be any experiment and let $\mathcal{B}$ be any subfield of $\mathcal{A}$. Then for any $\mu\in L(\mathcal{E}(\mathcal{B}))$ there exists $\nu\in L(\mathcal{E})$ such that $\nu/\mathcal{B}=\mu$.

<u>Proof</u>. By Lemma 5.1 there exists a countable subfamily

$\mathcal{P}_0=\{P_{\theta_1},P_{\theta_2},\ldots\}$ of $\mathcal{P}$ such that μ belongs to the band generated by $\mathcal{P}_0/\mathcal{B}$ in $\mathfrak{M}(\mathcal{B})$. Since

$$n(B)=\sum 2^{-i}(P_{\theta_i}/\mathcal{B})(B)$$

is an equivalent dominating measure for the dominated experiment $(\mathcal{X},\mathcal{B},\mathcal{P}_0/\mathcal{B})$ there exists an $f\in L^1(\mathcal{X},\mathcal{B},n)$ such that $\mu = f\cdot n$ by Theorem 5.2. Let n' be the measure on $(\mathcal{X},\mathcal{A})$ defined by "lifting" n to $\mathcal{A}$, that is

$$n'(A)=\sum 2^{-i}P_{\theta_i}(A),\ A\in\mathcal{A}.$$

Then $\nu=f\cdot n'$ belongs to the band generated by $\mathcal{P}_0$ in $\mathfrak{M}(\mathcal{A})$ by Theorem 5.2 again. Hence $\nu\in L(\mathcal{E})$ and $\nu/\mathcal{B}=\mu$.

<u>Theorem 5.3</u>. Let $\mathcal{E}=(\mathcal{X},\mathcal{A},\mathcal{P}=\{P_\theta;\theta\in\Theta\})$ be any experiment and $\mathcal{B}$ be pairwise sufficient for $\mathcal{E}$. Then $\delta(\mathcal{E}(\mathcal{B}),\mathcal{E})=0$.

<u>Proof</u>. For any $\mu\in L(\mathcal{E}(\mathcal{B}))$, by Lemma 5.2, there exists $\mu'\in L(\mathcal{E})$ such that $\mu=\mu'/\mathcal{B}$. Then there exists a countable subfamily $\mathcal{P}_0$ of $\mathcal{P}$ such that μ' belongs to the band generated by $\mathcal{P}_0$ in $\mathfrak{M}(\mathcal{A})$ by Lemma 5.1. Define a mapping T on $L(\mathcal{E}(\mathcal{B}))$ into $L(\mathcal{E})$ by

$$T\mu(A)=\int E(I_A/\mathcal{B})\,d\mu,\ A\in\mathcal{A}.$$

Here $E(I_A/\mathcal{B})$ is the conditional probability of A given $\mathcal{B}$ commmon to all P in $\mathcal{P}_0$ whose existence is guaranteed by sufficiency of $\mathcal{B}$ for $(\mathcal{X},\mathcal{A},\mathcal{P}_0)$. Then it is easily proved that T is well defined. T is a transition from $\mathcal{E}(\mathcal{B})$ to $\mathcal{E}$ because $T\mu\ll\mathcal{P}_0$. T satisfies that $TP_\theta/\mathcal{B}=P_\theta$ for all $\theta\in\Theta$. These imply $\delta(\mathcal{E}(\mathcal{B}),\mathcal{E})=0$.

It is known that the converse of Theorem 5.3 is also true.This follows, for example, from 19.1, 53.8 and 54.6 of Strasser(1985).

By Theorem 5.3 $\delta(\mathcal{E}(\mathcal{B}),\mathcal{E})=0$ for any PSS subfield $\mathcal{B}$ for a majorized experiment $\mathcal{E}$. In the proof of Theorem 5.3 we constructed a transition which makes the deficiency zero. For any majorized experiment $\mathcal{E}$, if $\mathcal{B}$ is PSS for $\mathcal{E}$ then using a pivotal measure we can construct a transition from $\mathcal{E}(\mathcal{B})$ to $\mathcal{E}$

which makes the deficiency zero.

Lemma 5.3. Let $\mathcal{E}$ be a majorized experiment and $\mathcal{B}$ be a PSS subfield of $\mathcal{A}$. Then for any subfield $\mathcal{C}\subset\mathcal{B}$, $\mathcal{C}$ is PSS for $\mathcal{E}(\mathcal{B})$ if and only if $\mathcal{C}$ is PSS for $\mathcal{E}$.

The proof is easy, so it is omitted.

Lemma 5.4. Let $\mathcal{E}$ be a majorized experiment and $\mathcal{B}$ be a PSS subfield of $\mathcal{A}$. Then, for any pivotal measure n for $\mathcal{E}$, $n/\mathcal{B}$ is a pivotal measure for $\mathcal{E}(\mathcal{B})$.

Proof. That $n/\mathcal{B}$ is an equivalent majorizing measure for $\mathcal{E}(\mathcal{B})$ is clear. So let $\mathcal{C}(\subset\mathcal{B})$ be any PSS subfield for $\mathcal{E}(\mathcal{B})$. Then $\mathcal{C}$ is PSS for $\mathcal{E}$ by Lemma 5.3. Therefore any P_θ in $\mathcal{P}$ has a $\mathcal{C}$-measurable version of the density w.r.t. n and it is also a version of the density of $P_\theta/\mathcal{B}$ w.r.t. $n/\mathcal{B}$.

Conversely let any $P_\theta/\mathcal{B}$ have a $\mathcal{C}$-measurable version of the density w.r.t. $n/\mathcal{B}$. Then, by the Neyman factorization theorem (Theorem 2.3) for the majorized experiment $\mathcal{E}(\mathcal{B})$, $\mathcal{C}$ is PSS for $\mathcal{E}(\mathcal{B})$. Hence $n/\mathcal{B}$ is a pivotal measure for $\mathcal{E}(\mathcal{B})$.

Let $\mathcal{E}=(\mathcal{X},\mathcal{A},\mathcal{P}=\{P_\theta;\theta\in\Theta\})$ be a majorized experiment and let $\mathcal{B}$ be any PSS subfield of $\mathcal{A}$. Then by Lemma 5.4, for any pivotal measure n for $\mathcal{E}$, $n/\mathcal{B}$ is a pivotal measure for $\mathcal{E}(\mathcal{B})$. Then, by Theorem 5.2, we have

$$L(\mathcal{E})=\{f\cdot n;f\in L^1(\mathcal{X},\mathcal{A},n)\}$$

and

$$L(\mathcal{E}(\mathcal{B}))=\{f\cdot n/\mathcal{B};f\in L^1(\mathcal{X},\mathcal{B},n/\mathcal{B})\}.$$

Then the mapping T_n defined by

$$T_n(f\cdot n/\mathcal{B})=f\cdot n,\quad f\in L^1(\mathcal{X},\mathcal{B},n/\mathcal{B}),$$

is a transition from $\mathcal{E}(\mathcal{B})$ to $\mathcal{E}$. Each P_θ has a $\mathcal{B}$-measurable density w.r.t. n, and hence T_n maps $P_\theta/\mathcal{B}$ to P_θ for all $\theta\in\Theta$. Therefore we have

$$sup\{\|T_nP_\theta/\mathcal{B}-P_\theta\|;\theta\in\Theta\}=0,$$

which implies

$$\delta(\mathcal{E}(\mathcal{B}),\mathcal{E})=inf\{sup\{\|TP_\Theta/\mathcal{B}-P_\Theta\|;\theta\in\Theta\};T\}=0.$$

Thus T_n is a transition which makes the deficiency of $\mathcal{E}(\mathcal{B})$ relative to $\mathcal{E}$ zero.

5.2. Representation of *M*-space of majorized experiment

In this section, using a pivotal measure, we discuss a relation between two representations of the *M*-space of a majorized experiment. Suppose that $\mathcal{E}=(\mathcal{X},\mathcal{A},\mathcal{P}=\{P_\theta;\theta\in\Theta\})$ is a majorized experiment and n is an equivalent majorizing measure for $\mathcal{E}$. Then, by Theorem 5.2, $L(\mathcal{E})=\{f\cdot n;f\in L^1(\mathcal{X},\mathcal{A},n)\}$. So $L(\mathcal{E})$ is isometrically isomorphic to $L^1(\mathcal{X},\mathcal{A},n)$.

McShane (1962) studied the representation of the dual space of $L^1(\mathcal{X},\mathcal{A},m)$ for a general measure m which is not necessarily σ-finite. Hence such a representation naturally relates to the representation of the *M*-space of a majorized experiment.

Let $(\mathcal{X},\mathcal{A},m)$ be any measure space and define $m(\mathcal{A})=\{m/E;E\in\mathcal{A}(m)\}$.

Definition 5.2. A family

$$b(\mathcal{A}(m))=\{b(x,E);E\in\mathcal{A}(m)\}$$

of $\mathcal{A}$-measurable real valued function on $\mathcal{X}$ is called a (measurable) *quasi-function* w.r.t. $m(\mathcal{A})$ if for any countable subfamily $\{E_1,E_2,...\}$ of $\mathcal{A}(m)$ there exists an $\mathcal{A}$-measurable function $b(\cdot)$ such that $b(x)=b(x,E_i)$ a.e. m/E_i for all $i\geq 1$. A quasi-function $b(\mathcal{A}(m))$ w.r.t. $m(\mathcal{A})$ is said to be *essentially bounded* if there exists a real number r such that $|b(x,E)|\leq r$ a.e. m/E for all $E\in\mathcal{A}(m)$. Then the infimum of such r's is called the *essential supremum* of $b(\mathcal{A}(m))$ and is denoted, as it is considered as a norm, by

$$\|b(\mathcal{A}(m))\|=ess.sup|b(\mathcal{A}(m))|.$$

Remark 5.1. If $b(\mathcal{A}(m))=\{b(x,E);E\in\mathcal{A}(m)\}$ is a quasi-function w.r.t. $m(\mathcal{A})$, then w.l.g. we can assume that $b(x,E)=0$ for all $x\in\mathcal{X}-E$. Any family $\{b(x,E);E\in\mathcal{A}(m)\}$ with this property is a quasi-function w.r.t. $m(\mathcal{A})$ if and only if this family is an m-cross section. We omit the detail(*cf*. Rao(1975),section

2.2).

Definition 5.3. The *(M)-integral* of a quasi-function $b(\mathcal{A}(m))$ w.r.t. $m(\mathcal{A})$ is denoted by

$$(M)-\int b(\mathcal{A}(m))\ dm,$$

and is defined as the extended real number k, if it exists, such that for any neighbourhood $U = U(k)$ of k there exists a set $E(U)\in\mathcal{A}(m)$ satisfying

$$\int b(x,E)\,dm/E \in U$$

for all $E\supset E(U),E\in\mathcal{A}(m)$.

For any $f\in L^1(\mathcal{X},\mathcal{A},m)$ the M-integral w.r.t. $f\cdot m$ of a quasi-function $b(\mathcal{A}(m))$ w.r.t. $m(\mathcal{A})$, denoted by

$$(M)-\int b(\mathcal{A}(m))\ d(f\cdot m),$$

is defined as the (M)-integral of the quasi-function $b(\mathcal{A}(m))\cdot f=\{b(x,M)f(x);E\in\mathcal{A}(m)\}$ w.r.t. $m(\mathcal{A})$, that is

$$(M)-\int b(\mathcal{A}(m))\ d(f\cdot m) = (M)-\int b(\mathcal{A}(m))\cdot f\ dm.$$

Theorem 5.4. (Riesz's theorem) Let $(\mathcal{X},\mathcal{A},m)$ be a measure space with the finite subset property. Then for any continuous linear functional l on $L^1(\mathcal{X},\mathcal{A},m)$ there exists an essentially bounded $\mathcal{A}$-measurable quasi-function $b(\mathcal{A}(m))$ w.r.t. $m(\mathcal{A})$ such that

$$l(f) = (M)-\int b(\mathcal{A}(m))\ d(f\cdot m), \qquad f\in L^1(\mathcal{X},\mathcal{A},m),$$

and $\|l\|=\|b(\mathcal{A}(m))\|$. The $b(\mathcal{A}(m))$ is uniquely determined up to equivalence.

This is due to McShane (1962) and the proof is omitted.

Remark 5.2. McShane showed this theorem for quasi-functions w.r.t. the family of measures corresponding to localizable sets under the assumption that each locally measurable set is measurable. But on perusal of his proof one finds that the theorem can also be stated as in Theorem 5.4.

Again let $\mathcal{E}=(\mathcal{X},\mathcal{A},\mathcal{P}=\{P_\theta;\theta\in\Theta\})$ be a majorized experiment and let m be an equivalent majorizing measure for $\mathcal{E}$. By Theorem 5.4, $\mathbf{M}(\mathcal{E})$ is the totality of essentially bounded $\mathcal{A}$-measurable quasi-functions $\{b(\mathcal{A}(m))\}$ w.r.t. $m(\mathcal{A})$. And for any $f\cdot m\in L(\mathcal{E})$ we have

$$b(\mathcal{A}(m))(f\cdot m)=(\mathbf{M})-\int b(\mathcal{A}(m))\ d(f\cdot m).$$

If m is a localizable measure, that is if $\mathcal{E}$ is weakly dominated, then for any such $b(\mathcal{A}(m))$ there exists a bounded $\mathcal{A}$-measurable function b such that $b(x)=b(x,E)$ a.e. m on E, and

$$(\mathbf{M})-\int b(\mathcal{A}(m))\ d(f\cdot m)\ =\ \int b(x)\ d(f\cdot m).$$

In particular the $\mathbf{M}$-space of a weakly dominated experiment is isometrically isomorphic to the space of bounded $\mathcal{A}$-measurable functions. This remark follows from Theorem 1.3 and Remark 5.1.

On the other hand Torgersen (1979,1991(Theorem 1.2.16)) (also refer to Bomze (1990)) characterized $\mathbf{M}(\mathcal{E})$ for *any* experiment $\mathcal{E}$ by introducing an integral. Let's call it *(T)-integral*. Let $\mathcal{E}=(\mathcal{X},\mathcal{A},\mathcal{P}=\{P_\theta;\theta\in\Theta\})$ be any experiment. Then $\mathbf{M}(\mathcal{E})$ is the totality of essentially bounded countably coherent families $\{f(\mathcal{P})=\{f(x,\theta);\theta\in\Theta\}\}$ of $\mathcal{A}$-measurable functions, and it holds

$$f(\mathcal{P})(\mu)=\ (T)-\int f(\mathcal{P})\ d\mu,\qquad \mu\in L(\mathcal{E}).$$

Here the right hand side is the (T)-integral, which is defined as follows. By Lemma 5.1 for any $\mu\in L(\mathcal{E})$ there exists a countable subfamily $\{\theta_1,\theta_2,\ldots\}$ of Θ such that μ belongs to the band generated by $\{P_{\theta_1},P_{\theta_2},\ldots\}$ in $\mathfrak{M}(\mathcal{A})$. Then there exists an $\mathcal{A}$-measurable bounded function $f(\cdot)$ such that $f(x)=f(x,\theta_i)$ a.e. P_{θ_i} for all $i\geq 1$.

<u>Definition 5.4</u>. The *(T)-integral* of $f(\mathcal{P})$ w.r.t. μ is

defined as $\int f(x)\ d\mu$; that is,

$$(T)-\int f(P)\ d\mu = \int f(x)\ d\mu.$$

Therefore we have two representations of the M-space of a majorized experiment. So it is clear that for any (M)-integral representation there corresponds a (T)-integral representation and vice versa. Hence the problem to investigate the correspondence between them naturally arises. (M)-integral and (T)-integral are defined in terms of cross-section property and finitely (countably) coherent property of a family of functions respectively. These properties were already given in Section 1.2, Definition 1.5 and Definition 1.7 respectively. But in that section we considered the former as a property of a measure and the latter of an experiment. By combining them we answer the above problem in Theorem 5.5. For $\mu\in\mathfrak{M}(\mathcal{A})_+$, if there exists a set $S\in\mathcal{A}$ such that (1) $\mu(\mathcal{X}-S)=0$, and (2) $A\subset S$, $\mu(A)=0$ implies that $A\in N(\mathcal{P})$, then S is said to be a *support* for $\mathcal{E}$ of μ. Of course this definition coincides with the one previously defined in Definition 1.2 in the case that $\mu\in\mathcal{P}$. Let us denote

$$U(\mathcal{E})=\{\mu\in\mathfrak{M}(\mathcal{A});\ |\mu| \text{ has a support for } \mathcal{E}\},$$

and

$$V(\mathcal{E})=\{\mu\in\mathfrak{M}(\mathcal{A});\ |\mu|\ll\mathcal{P}\}.$$

Next Lemma gives properties of L-space of an experiment.

Lemma 5.5. Let $\mathcal{E}=(\mathcal{X},\mathcal{A},\mathcal{P}=\{P_\theta;\theta\in\Theta\})$ be any experiment. Then

(1) $U(\mathcal{E})\cap V(\mathcal{E})\subset L(\mathcal{E})\subset V(\mathcal{E})$,

(2) $\mathcal{E}$ is majorized if and only if $L(\mathcal{E})=U(\mathcal{E})\cap V(\mathcal{E})$.

Proof. (1) We can easily prove that $U(\mathcal{E})$ and $V(\mathcal{E})$ are bands. Hence $U(\mathcal{E})\cap V(\mathcal{E})$ is also a band in $\mathfrak{M}(\mathcal{A})$. We show

$$L(\mathcal{E})^\perp\cap U(\mathcal{E})\cap V(\mathcal{E})=\{0\},$$

where $L(\mathcal{E})^\perp$ denotes the totality of $\mu\in\mathfrak{M}(\mathcal{A})$ such that μ is orthogonal to each $\nu\in L(\mathcal{E})$, i.e. $|\mu|\wedge|\nu|=0$. Let $\mu\in L(\mathcal{E})^\perp\cap U(\mathcal{E})$ $\cap V(\mathcal{E})$ and let S be a support of $|\mu|$ for $\mathcal{E}$. Then we have $P_\theta(S\cap\cdot)\ll|\mu|$ and $P_\theta(S\cap\cdot)\wedge|\mu|\leqq P_\theta\wedge|\mu|=0$ for all $\theta\in\Theta$. This implies $P_\theta(S)=0$ for all $\theta\in\Theta$. Hence $|\mu|(S)=0$ because $\mu\in V(\mathcal{E})$, which implies $\mu=0$. By Riesz decomposition theorem (Schaefer(1974),

Chapter 2, Theorem 2.10), we have $U(\mathcal{E})\cap V(\mathcal{E})\subset L(\mathcal{E})$. Since $V(\mathcal{E})$ is a band including $\mathcal{P}$ it is clear that $L(\mathcal{E})\subset V(\mathcal{E})$.

(2) Suppose that $\mathcal{E}$ is a majorized experiment and let n be a majorizing measure for $\mathcal{E}$. Since each P_θ has a support for $\mathcal{E}$, $U(\mathcal{E})\cap V(\mathcal{E})$ is a band including $\mathcal{P}$, which implies $L(\mathcal{E})\subset U(\mathcal{E})\cap V(\mathcal{E})$. By (1) we have $L(\mathcal{E})=U(\mathcal{E})\cap V(\mathcal{E})$.

Conversely we assume that $L(\mathcal{E})=U(\mathcal{E})\cap V(\mathcal{E})$. Take any maximal orthogonal system $\{Q_t\}$ of $L(\mathcal{E})$ and let $\mu=\sum Q_t$. Then it follows that $\mathcal{P}\sim\mu$. By Assumption each Q_t has a support $S_t\in\mathcal{A}$. Then we have $\mu(S_t\cap\cdot)=Q_t(\cdot)$ for all t because of orthogonality of the system $\{Q_t\}$. Thus we have $\{Q_t\}\subset\{f\cdot\mu;f\in L^1(\mathcal{X},\mathcal{A},\mu)\}$. Since the right hand side is a band, and since $L(\mathcal{E})$ is the band generated by $\{Q_t\}$ we have $L(\mathcal{E})\subset\{f\cdot\mu;f\in L^1(\mathcal{X},\mathcal{A},\mu)\}$. This shows that μ is a majorizing measure of $\mathcal{E}$.

Let $\mathcal{E}=(\mathcal{X},\mathcal{A},\mathcal{P}=\{P_\theta;\theta\in\Theta\})$ be a majorized experiment. Suppose that $\mathcal{B}_0$ is the smallest PSS subfield. Let $S(\theta)$ be a support of P_θ for $\mathcal{E}$ which belongs to $\mathcal{B}_0$. It is also a support of $P_\theta/\mathcal{B}_0$ for $\mathcal{E}(\mathcal{B}_0)$. Let $\mathcal{F}=\{F\}$ be a maximal decomposition of the majorized experiment $\mathcal{E}(\mathcal{B}_0)$. For each $F\in\mathcal{F}$ let P_F be a measure in $\mathcal{P}$ such that $P_F/\mathcal{B}_0(F)>0$ and $F-S(P_F)\in N(\mathcal{P}/\mathcal{B}_0)$. Then, by Theorem 2.2, the measure n on $\mathcal{A}$ defined by

$$n(A)=\sum\{P_F(A\cap F);F\in\mathcal{F}\}$$

is a pivotal measure for $\mathcal{E}$. This pivotal measure has a feature that $n=P_F$ on F for each $F\in\mathcal{F}$. We also note that, by Lemma 1.4, $\mathcal{F}$ is a maximal decomposition of $(\mathcal{X},\mathcal{A},n)$.

Suppose that $f(\mathcal{P})=\{f(x,\theta);\theta\in\Theta\}$ is any essentially bounded countably coherent family for a majorized experiment $\mathcal{E}=(\mathcal{X},\mathcal{A},\mathcal{P}=\{P_\theta;\theta\in\Theta\})$ and n is the pivotal measure for $\mathcal{E}$ which is defined in the same way as in the previous paragraph. Take any $E\in\mathcal{A}(n)$. Since $\mathcal{F}$ is also a maximal decomposition of $(\mathcal{X},\mathcal{A},n)$ the set $\{F\in\mathcal{F};n(E\cap F)>0\}$ is countable, say $\{F_1,F_2,\ldots\}$, and $n(E-\bigcup_{i=1}^{\infty}(E\cap F_i))=0$ by Proposition 1.1. For each $i\geq 1$, there exists $\theta_i\in\Theta$ such that $P_{\theta_i}=P_{F_i}$. Let us denote $\mathcal{P}(E)=\{P_{\theta_i};i\geq 1\}$. Then there exists a bounded $\mathcal{A}$-measurable function $b(\cdot;\mathcal{P}(E))$ such that

$$b(x;\mathcal{P}(E))=f(x,\theta_i)\text{ a.e. }P_{\theta_i}\text{ for all }i\geq 1.$$

Define

$$b(x,E)=b(x;P(E)) \qquad \text{if } x\in E,$$
$$\qquad =0 \qquad \text{if } x\bar{\in}E.$$

Then we have the following theorem giving the correspondence of (T)- and (M)-integral.

<u>Theorem 5.5</u>. For any essentially bounded countably coherent family $f(P)=\{f(x,\theta);\theta\in\Theta\}$ the family $b(\mathcal{A}(n))=\{b(x,E);E\in\mathcal{A}(n)\}$ defined above is an essentially bounded $\mathcal{A}$-measurable quasi-function w.r.t. $n(\mathcal{A})$. For any $\mu=f\cdot n$, $f\in L^1(\mathcal{X},\mathcal{A},n)$, in $L(\mathcal{E})$ we have

$$(T)-\int f(P)\ d\mu = (M)-\int b(\mathcal{A}(n))\ d\mu.$$

<u>Proof</u>. *Step* 1. Take any $E_1,E_2\in\mathcal{A}(n)$. Let $\{F_1,F_2,\ldots\}=\{F\in\mathcal{F};n(E_1\cap E_2\cap F)>0\}$. Then we have

$$b(x;P(E_1))=f(x,\theta_i)=b(x;E_2) \text{ a.e. } P_{\theta_i} \text{ for all } i\geq 1, \text{ and}$$

$$n(E_1\cap E_2\cap[b(x;P(E_1))\neq b(x;P(E_2))])$$

$$=\sum_{i=1}^{\infty} n(E_1\cap E_2\cap[b(x;P(E_1))\neq b(x;P(E_2))]\cap F_i)$$

$$=\sum_{i=1}^{\infty} P_{\theta_i}(E_1\cap E_2\cap[b(x;P(E_1))\neq b(x;P(E_2))]\cap F_i) = 0.$$

Hence $b(\mathcal{A}(n))$ is an n-cross section, so it is an essentially bounded $\mathcal{A}$-measurable quasi-function w.r.t. $n(\mathcal{A})$.

Step 2. Let $\mu=f\cdot n$, $f\in L^1(\mathcal{X},\mathcal{A},n)$, be any element of $L(\mathcal{E})$. We can write $[|f(x)|>0]=\sum_{i=1}^{\infty} E_i$ (disjoint union), $n(E_i)<\infty$ for all $i\geq 1$. For any $i\geq 1$ the set $\{F\in\mathcal{F};n(E_i\cap F)>0\}$ is countable, say $\{F_{i_1},F_{i_2},\ldots\}$, and $n(E_i-\bigcup_{j=1}^{\infty}(E_i\cap F_{i_j}))=0$. Then there exists $\theta_{i_j}\in\Theta$ such that $P_{\theta_{i_j}}(F_{i_j})>0$, $F_{i_j}-S(\theta_{i_j})\in N(P/\mathcal{B}_0)$. We show that $|\mu|=|f|\cdot n$ is absolutely continuous w.r.t. $\{P_{\theta_{i_j}};i,j\geq 1\}$. Let us assume that $P_{\theta_{i_j}}(A)=0$ for all $i,j\geq 1$. Since

$$|\mu|(A)=\int_A|f(x)|\ dn = \sum_{i=1}^{\infty}\int_{A\cap E_i}|f(x)|\ dn,$$

it must be proved that $n(A\cap E_i)=0$ for all $i\geq 1$. For this purpose it is enough to show $P_\theta(A\cap E_i\cap F)=0$ for all $F\in\mathcal{F}$ and $\theta\in\Theta$ since $\mathcal{F}$ is a maximal decomposition and n is equivalent to $\mathcal{P}$. It is clear that $\{F\in\mathcal{F};\ n(A\cap E_i\cap F)>0\}$ is a subset of $\{F_{i_1},F_{i_2},\ldots\}$ and $P_{\theta_{i_j}}(A\cap E_i\cap F_{i_j})=0$ for all $i,j\geq 1$. Hence, by $A\cap E_i\cap F_{i_j}-S(\theta_{i_j})\in N(\mathcal{P})$, we have

$$A\cap E_i\cap F_{i_j}=[A\cap E_i\cap F_{i_j}\cap S(\theta_{i_j})]\cup[A\cap E_i\cap F_{i_j}\cap(\mathcal{X}-S(\theta_{i_j}))]\in N(\mathcal{P}).$$

So it follows that $P_\theta(A\cap E_i\cap F)=0$ for all $F\in\mathcal{F}$, $\theta\in\Theta$. By Lemma 5.5, $\mu=f\cdot n$ belongs to the band generated by $\{P_{\theta_{i_j}};i,j\geq 1\}$ in the space $\mathfrak{M}(\mathcal{A})$.

Step 3. For this countable family $\{f(x,\theta_{i_j});\ i,j\geq 1\}$ there exists a bounded $\mathcal{A}$-measurable function f_0 such that $f_0(x)=f(x,\theta_{i_j})$ a.e. $P_{\theta_{i_j}}$ for all $i,j\geq 1$. Then by the definition of (T)-integral we have

$$(T)-\int\{f(x,\theta);\theta\in\Theta\}\ d\mu = \int f_0(x)\ d\mu = \int f_0(x)f(x)\ dn, \qquad (5.1)$$

On the other hand,

$$(M)-\int\{b(x,E);E\in\mathcal{A}(n)\}d\mu =$$

$$(M)-\int\{b(x,E)f(x);E\in\mathcal{A}(n)\}dn. \qquad (5.2)$$

We shall show that (5.1) is equal to (5.2).

Case 3.1. $f(x)\geq 0$, $f(x,\theta)\geq 0$ for all $x\in\mathcal{X}$ and $\theta\in\Theta$.
In this case we can assume that $f_0(x)$ and $b(x,E)$ are non-negative for all $E\in\mathcal{A}(n)$. For the countable family $\{E_i;i\geq 1\}$ defined in Step 2 there exists a bounded $\mathcal{A}$-measurable function $b(x;\{E_i\})$ such that

$$b(x;\{E_i\})=b(x;E_i)\ \text{a.e.}\ n/E_i\ \text{for all}\ i\geq 1.$$

Then it follows that

$$\int b(x;\{E_i\})f(x)\ dn = \sum_{i=1}^{\infty}\int_{E_i} b(x;\{E_i\})f(x)\ dn$$

$$=\sum_{i=1}^{\infty}\sum_{j=1}^{\infty}\int_{E_i\cap F_{i_j}} b(x;\{E_i\})\ f(x)\ dn$$

$$=\sum_{i=1}^{\infty}\sum_{j=1}^{\infty}\int_{E_i\cap F_{i_j}} f(x,\theta_{i_j})\ f(x)\ dP_{\theta_{i_j}}$$

$$=\sum_{i=1}^{\infty}\sum_{j=1}^{\infty}\int_{E_i\cap F_{i_j}} f_0(x)f(x)\ dP_{\theta_{i_j}}$$

$$=\sum_{i=1}^{\infty}\sum_{j=1}^{\infty}\int_{E_i\cap F_{i_j}} f_0(x)f(x)\ dn = \sum_{i=1}^{\infty}\int_{E_i} f_0(x)f(x)\ dn$$

$$=\int f_0(x)f(x)\ dn = (T)-\int\{f(x,\theta);\theta\in\Theta\}\ d\mu. \tag{5.3}$$

Here we used the fact that $n=P_F$ on F, and $b(x,E_i)=f(x,\theta_{i_j})$ a.e. $P_{\theta_{i_j}}$ on E_i. Since

$$|\int b(x;\{E_i\})f(x)\ dn| \le \int |b(x;\{E_i\})f(x)|\ dn<\infty, \text{ and}$$

$$\int b(x;\{E_i\})f(x)\ dn = \sum_{i=1}^{\infty}\int_{E_i} b(x;\{E_i\})f(x)\ dn\ ,$$

for any $\varepsilon>0$ there exists a natural number N_0 such that

$$\int b(x;\{E_i\})f(x)\ dn - \sum_{i=1}^{N_0}\int_{E_i} b(x;\{E_i\})f(x)\ dn <\varepsilon.$$

Then we have, for $E_0\equiv\sum_{i=1}^{N_0} E_i\in\mathcal{A}(n)$,

$$|\int b(x;\{E_i\})f(x)\ dn - \int b(x,E_0)f(x)\ dn|$$

$$=|\int b(x;\{E_i\})f(x)\ dn - \int_{E_0} b(x;\{E_i\})f(x)\ dn|<\varepsilon.$$

For any $E\in\mathcal{A}(n)$ such that $E\supset E_0$ we have

$$\int_{E-E_0} b(x,E-E_0)f(x)\ dn = \sum_{i=1}^{\infty}\int_{(E-E_0)\cap E_i} b(x;E-E_0)f(x)\ dn$$

$$=\sum_{i=N_0+1}^{\infty}\int_{(E-E_0)\cap E_i} b(x,E-E_0)f(x)\ dn$$

$$=\sum_{i=1}^{\infty}\int_{E\cap E_i} b(x,E_i)f(x)\ dn \leq \sum_{i=N_0+1}^{\infty}\int_{E_i} b(x,E_i)f(x)\ dn$$

$$=\sum_{i=N_0+1}^{\infty}\int_{E_i} b(x;\{E_i\})f(x)\ dn$$

$$=\int b(x;\{E_i\})f(x)\ dn - \sum_{i=1}^{N_0}\int_{E_i} b(x;\{E_i\})f(x)\ dn < \varepsilon.$$

Hence it follows that

$$|\int b(x;\{E_i\})f(x)\ dn - \int b(x,E)f(x)\ dn/_E|$$

$$=|\int b(x;\{E_i\})f(x)\ dn - \int b(x,E)f(x)\ dn/_{E_0} -$$

$$\int b(x,E)f(x)dn/_{E-E_0}|$$

$$=|\int b(x;\{E_i\})f(x)dn - \int b(x,E_0)f(x)dn/_{E_0} -$$

$$\int b(x,E-E_0)f(x)dn/_{E-E_0}|$$

$$\leq|\int b(x;\{E_i\})f(x)\ dn - \int b(x,E_0)f(x)\ dn/_{E_0}|$$

$$+\int b(x,E-E_0)f(x)dn/_{E-E_0} < \varepsilon+\varepsilon=2\varepsilon. \qquad (5.4)$$

Therefore, by the definition of the (M)-integral and (5.3), (5.4), we have

$$(M)-\int\{b(x,E);E\in\mathcal{A}(n)\}\ d\mu$$

$$=(M)-\int\{b(x,E)f(x);E\in\mathcal{A}(n)\}\ dn$$

$$=\int b(x;\{E_i\})f(x)\ dn = (T)-\int\{f(x,\theta);\theta\in\Theta\}\ d\mu.$$

Case 3.2. Only $f(x,\theta)\geq 0$ for all $x\in\mathcal{X},\theta\in\Theta$ are assumed. In this case we decompose any $f\in L^1(\mathcal{X},\mathcal{A},n)$ as $f=f^+-f^-, f^+, f^-\in L^1(\mathcal{X},\mathcal{A},n)$. Then we have $f\cdot n=f^+\cdot n - f^-\cdot n$. To $f^+\cdot n$ and $f^-\cdot n$ we can apply the previous results. So we omit the details. But here we need to note that the quasi-function $\{b(x,E);E\in\mathcal{A}(n)\}$ could be constructed depending only on n and independently of μ or f, and that

$$(M)-\int\{b(x,E)f^+(x);E\in\mathcal{A}(n)\}\ dn\ -\ (M)-\int\{b(x,E)f^-(x);$$

$$E\in\mathcal{A}(n)\}\ dn = (M)-\int\{b(x,E)f(x);E\in\mathcal{A}(n)\}\ dn.$$

Case 3.3. General case.
In this case we divide $f(x,\theta)$ as $f(x,\theta) = f^+(x,\theta)-f^-(x,\theta)$. Then it is clear that $\{f^+(x,\theta);\theta\in\Theta\}$ and $\{f^-(x,\theta);\theta\in\Theta\}$ are non-negative essentially bounded countably coherent families. Let $\{b_+(x,E);E\in\mathcal{A}(n)\}$, $\{b_-(x,E);E\in\mathcal{A}(n)\}$ be the corresponding essentially bounded measurable quasi-functions w.r.t. $n(\mathcal{A})$. Let $f_{0,+}(x)$ (resp. $f_{0,-}(x)$) be the bounded $\mathcal{A}$-measurable function which appears in Step 3 in the definition of the (T)-integral of $\{f^+(x,\theta);\theta\in\Theta\}$ (resp. $\{f^-(x,\theta);\theta\in\Theta\}$). Then it follows that

$$(T)-\int\{f^{\pm}(x,\theta);\theta\in\Theta\}f(x)\ dn = \int f_{0,\pm}(x)f(x)\ dn$$

$$=(M)-\int\{b_{\pm}(x,E);E\in\mathcal{A}(n)\}f(x)\ dn.$$

Here we note that the functions $f_{0,\pm}$ can be determined by $\{f^{\pm}(x,\theta_{i_j});i,j\geq 1\}$ through the same family of parameters $\{\theta_{i_j};i,j\geq 1\}$. Hence we have

$$(T)-\int\{f(x,\theta);\theta\in\Theta\}\ f(x)\ dn = \int f_0(x)f(x)\ dn,$$

where $f_0(x)=f_{0,+}(x) - f_{0,-}(x)$. Then it is easy to show that

$$(T)-\int\{f(x,\theta);\theta\in\Theta\}\ d\mu = (M)-\int\{b(x,E);E\in\mathcal{A}(n)\}\ d\mu.$$

This completes the proof.

If we start from a quasi-function we have another

correspondence of (M)- and (T)-integral. Let $\mathcal{E}=(\mathcal{X},\mathcal{A},\mathcal{P}=\{P_\theta;\theta\in\Theta\})$ be a majorized experiment and n be the pivotal measure for $\mathcal{E}$ defined in the paragraph before Theorem 5.5. Let $\{b(x,E);E\in\mathcal{A}(n)\}$ be any essentially bounded $\mathcal{A}$-measurable quasi-function w.r.t. $n(\mathcal{A})$. Take any $\theta\in\Theta$ and let $[dP_\theta/dn(x)>0]=\sum_{i=1}^{\infty}E_i$, $n(E_i)<\infty$, $i\geq1$. Then for this $\{E_i;i\geq1\}$ there exists an $\mathcal{A}$-measurable bounded function $f(x,\theta)$ such that

$$f(x,\theta) = b(x,E_i) \text{ a.e. } n/_{E_i}, \text{ for all } i\geq1.$$

Then we have the following:

Theorem 5.6. For any essentially bounded $\mathcal{A}$-measurable quasi-function $b(\mathcal{A}(n)) = \{b(x,E);E\in\mathcal{A}(n)\}$ the family $f(\mathcal{P})=\{f(x,\theta);\theta\in\Theta\}$ defined above is an essentially bounded countably coherent family. For any $\mu\in L(\mathcal{E})$ we have

$$(M)-\int b(\mathcal{A}(n))\ d\mu = (T)-\int f(\mathcal{P})\ d\mu.$$

The proof is quite similar to that of Theorem 5.5. So we omit the proof.

Remark 5.3. The composition of the correspondences in Theorem 5.5 and Theorem 5.6 induces the identity.

Next theorem characterizes weakly dominated experiments by the M-spaces.

Theorem 5.7. An experiment $\mathcal{E}$ is weakly dominated if and only if $M(\mathcal{E})$ is isomorphic to the space of all bounded $\mathcal{A}$-measurable functions, that is,

$$\langle f,\mu\rangle=\int f(x)\ d\mu,\ \mu\in L(\mathcal{E}) \text{ and } f\in M(\mathcal{E}).$$

Proof. The "only if" part follows from the discussion in the paragraph after Remark 5.2.

"If" part. In this case any essentially bounded countably coherent family is determined by an $\mathcal{A}$-measurable bounded function. Hence, by Theorem 1.5, $\mathcal{E}$ is weakly dominated.

5.3. Representation of the common conditional probability for PSS by (T)-integral

In section 5.1 we proved that the transition

$$T_n:\quad L(\mathcal{E}(\mathcal{B}))\to L(\mathcal{E})$$
$$f\cdot n/\mathcal{B}\to f\cdot n,\quad f\in L^1(\mathcal{X},\mathcal{B},n/\mathcal{B})$$

makes the deficiency of $\mathcal{E}(\mathcal{B})$ relative to $\mathcal{E}$ zero, where n is any pivotal measure and $\mathcal{B}$ is any PSS subfield of a majorized experiment $\mathcal{E}=(\mathcal{X},\mathcal{A},P=\{P_\theta;\theta\in\Theta\})$. In this section we show that the dual mapping of T_n has the characteristic property which is appropriate to be called a common conditional probability given $\mathcal{B}$ though $\mathcal{B}$ is simply PSS and is not sufficient.

Though the notion of common conditional probability was defined in Definition 2.1, we will repeat here the definition because of its importance.

DEFINITION. Let $\mathcal{E}=(\mathcal{X},\mathcal{A},P=\{P_\theta;\theta\in\Theta\})$ be an experiment and $\mathcal{B}$ be sufficient for $\mathcal{E}$. Then for any $A\in\mathcal{A}$ there exists a $\mathcal{B}$-measurable function $E(I_A/\mathcal{B})$ such that

$$\int_B E(I_A/\mathcal{B})(x)\ dP_\theta/\mathcal{B} = P_\theta(A\cap B)\ \text{ a.e. } P_\theta \text{ for all } \theta.$$

Such a function $E(I_A/\mathcal{B})$, which is uniquely determined w.r.t. a.e. P_θ for all θ, is called the *common conditional probability* of A given $\mathcal{B}$.

Let us define a mapping ρ which denotes the canonical embedding of the space $\mathcal{F}(\mathcal{A})$ of $\mathcal{A}$-measurable bounded real valued functions on $\mathcal{X}$ into the M-space of $\mathcal{E}$. Namely for any $f\in\mathcal{F}(\mathcal{A})$ we define $\rho(f)=\{f;\theta\in\Theta\}$. The right hand side is the essentially bounded countably coherent family $\{f(x,\theta);\theta\in\Theta\}$ such that $f(x,\theta)=f(x)$ a.e. P_θ for all $\theta\in\Theta$. Similarly $\rho_{\mathcal{B}}$ denotes the embedding of $\mathcal{F}(\mathcal{B})$ into $M(\mathcal{E}(\mathcal{B}))$ where $\mathcal{B}$ is a PSS subfield for a majorized experiment $\mathcal{E}$.

Let us consider the dual mapping T_n' of T_n;

$$T_n':M(\mathcal{E})\longrightarrow M(\mathcal{E}(\mathcal{B})),$$
$$\langle T_n' f,\ \mu\rangle = \langle f,\ T_n\mu\rangle,\quad f\in M(\mathcal{E}),\quad \mu\in L(\mathcal{E}(\mathcal{B})).$$

Then we have

<u>Theorem 5.8</u>. T_n' satisfies the following:

(1) $(T)-\int T_n'\rho(I_A)\ d(I_B\cdot n/\mathcal{B})=n(A\cap B)$, $A\in\mathcal{A}, B\in\mathcal{B}(n/\mathcal{B})$; (5.5)

(2) $(T)-\int T_n'\rho(I_A)\ d(I_B\cdot P_\theta/\mathcal{B}) = P_\theta(A\cap B)$, $A\in\mathcal{A}, B\in\mathcal{B}, \theta\in\Theta$; (5.6)

(3) $T_n'\rho(I_A)=\{E(I_A/\mathcal{B},P_\theta);\theta\in\Theta\}$, $A\in\mathcal{A}$. (5.7)

<u>Proof</u>. It is easy to prove that T_n' is well defined. Let $\{f(x,\theta);\theta\in\Theta\}$ be the representation of $T_n'\rho(I_A)$ as a $\mathcal{B}$-measurable essentially bounded countably coherent family. For any $f\cdot n/\mathcal{B}\in L(\mathcal{E}(\mathcal{B}))$, $f\in L^1(\mathcal{X},\mathcal{B},n/\mathcal{B})$, we have

$$T_n'\rho(I_A)(f\cdot n/\mathcal{B})=\rho(I_A)(T_nf\cdot n/\mathcal{B})=\rho(I_A)(f\cdot n)$$

$$=(T)-\int\rho(I_A)f\ dn = \int I_Af\ dn.$$

Hence, for $f=I_B$, $B\in\mathcal{B}(n/\mathcal{B})$, we have

$$(T)-\int T_n'\rho(I_A)I_B\ dn/\mathcal{B} = n(A\cap B).$$

This proves (5.5). Take any $P_{\theta_0}\in\mathcal{P}$ and let f_0 be a $\mathcal{B}$-measurable version of the density of P_{θ_0} w.r.t. n. This is clearly a version of the density of $P_{\theta_0}/\mathcal{B}$ w.r.t. $n/\mathcal{B}$. Then, for any $B\in\mathcal{B}$, $I_Bf_0\in L^1(\mathcal{X},\mathcal{B},n/\mathcal{B})$. And so we have

$$(T)-\int\{f(x,\theta);\theta\in\Theta\}\ d(I_Bf_0\cdot n/\mathcal{B}) = T_n'\rho(I_A)(I_Bf_0\cdot n/\mathcal{B})$$

$$=\rho(I_A)(T_nI_Bf_0\cdot n/\mathcal{B})=\rho(I_A)(I_Bf_0\cdot n)=(T)-\int\rho(I_A)I_Bf_0\ dn$$

$$=P_{\theta_0}(A\cap B). \tag{5.8}$$

This proves (5.6). On the other hand $I_Bf_0\cdot n/\mathcal{B}(\in L(\mathcal{E}(\mathcal{B})))$ is absolutely continuous w.r.t. $P_{\theta_0}/\mathcal{B}$. Hence, by Lemma 5.5, $I_Bf_0\cdot n/\mathcal{B}$ belongs to the band generated by $P_{\theta_0}/\mathcal{B}$ in the space $\mathfrak{M}(\mathcal{B})$. Therefore by the definition of (T)-integral we have

$$(T)-\int\{f(x,\theta);\theta\in\Theta\}\ d(I_Bf_0\cdot n/\mathcal{B}) = \int f(x,\theta_0)\ d(I_Bf_0\cdot n/\mathcal{B})$$

$$=\int f(x,\theta_0) I_B \, dP_{\theta_0}/\mathcal{B}. \tag{5.9}$$

From (5.8) and (5.9) we have

$$\int_B E(I_A/\mathcal{B},P_{\theta_0}) \, dP_{\theta_0}/\mathcal{B} = P_{\theta_0}(A\cap B) = \int_B f(x,\theta_0) \, dP_{\theta_0}/\mathcal{B}$$

for all $B\in\mathcal{B}$. Hence it follows that $f(x,\theta_0)=E(I_A/\mathcal{B},P_{\theta_0})(x)$ a.e. P_{θ_0}. Since P_{θ_0} is arbitrary we have

$$T_n'\rho(I_A) = \{f(x,\theta);\theta\in\Theta\} = \{E(I_A/\mathcal{B},P_\theta);\theta\in\Theta\}.$$

This proves (5.7).

Though (T)-integral is used, (2) of Theorem 5.8 coincides with the definition of the common conditional probability in the DEFINITION. Thus T_n' is, in an extended sense, the common conditional probability and at the same time is the conditional probability of A w.r.t. n given $\mathcal{B}$ by (1) of the above theorem. Here we note that (3) of Theorem 5.8, which seems to be trivial, must be proved. If $\mathcal{B}$ is sufficient, by the above theorem,

$$T_n'\rho(I_A) = E(I_A/\mathcal{B}) = E(I_A/\mathcal{B},n) \text{ a.e. } n.$$

This corresponds to Theorem 5.1 in which dominated experiment case is stated. Thus PSS subfield $\mathcal{B}$ for a majorized experiment is "sufficient w.r.t. (T)-integral", and the corresponding common conditional probability is given by $T_n'\rho(I_A)$.

<u>Example 5.1</u>. Let $\mathcal{X}=R^1$(real line), and $\mathcal{A}=2^{\mathcal{X}}$. Let $\mathcal{P}$ be the totality of point probability measures. Then $\mathcal{E}=(\mathcal{X},\mathcal{A},\mathcal{P})$ is weakly dominated and the counting measure n is a pivotal measure for $\mathcal{E}$. Let $\mathcal{B}$ be the Borel field. Then $\mathcal{B}$ is PSS, but is not sufficient. $E(I_A/\mathcal{B},n)$ does not exist for any non-Borel set A. Our $T_n'\rho(I_A)$ exists in this case and is equal to

$$\{\{1;P_x\in\mathcal{P},x\in A\},\{0;P_x\in\mathcal{P},x\bar{\in} A\}\},$$

where P_x is the point probability measure at x.

<u>Corollary 5.2</u>. Let $\mathcal{E}=(\mathcal{X},\mathcal{A},\mathcal{P})$ be a majorized experiment and $\mathcal{B}$ be sufficient for $\mathcal{E}$. If n is a pivotal measure for $\mathcal{E}$ then we have $T_n'\rho(I_A) = \rho_{\mathcal{B}}(E(I_A/\mathcal{B})), A\in\mathcal{A}$.

<u>Remark 5.4</u>. In case $\mathcal{B}$ is sufficient, from Theorem 5.8 and Corollary 5.2, we have the following representation of T_n:

$$T_n(f\cdot n/\mathcal{B})(A)=\int E(I_A/\mathcal{B})(x)\; d(f\cdot n/\mathcal{B}), A\in\mathcal{A}, f\in L^1(\mathcal{X},\mathcal{B},n/\mathcal{B}).$$

Corollary 5.2 shows that, if $\mathcal{B}$ is sufficient, the T_n' transforms a bounded $\mathcal{A}$-measurable function to a bounded $\mathcal{B}$-measurable function. We prove that the converse is also true.

<u>Theorem 5.9</u>. Let $\mathcal{E}=(\mathcal{X},\mathcal{A},\mathcal{P}=\{P_\theta;\theta\in\Theta\})$ be a majorized experiment and $\mathcal{B}$ be PSS for $\mathcal{E}$. Let n be any pivotal measure for $\mathcal{E}$. If $T_n'\rho(\mathcal{F}(\mathcal{A}))\subset\rho_{\mathcal{B}}(\mathcal{F}(\mathcal{B}))$ then $\mathcal{B}$ is sufficient for $\mathcal{E}$ and $T_n'\rho(I_A)=\rho_{\mathcal{B}}(E(I_A/\mathcal{B})), A\in\mathcal{A}$.

<u>Proof</u>. For any $A\in\mathcal{A}$, by the assumption, there exists $g_A\in\mathcal{F}(\mathcal{B})$ such that $T_n'\rho(I_A)=\rho_{\mathcal{B}}(g_A)$. Take any $\theta\in\Theta$ and $B\in\mathcal{B}$. Let $S(\theta)$ be a support of P_θ for $\mathcal{E}$ belonging to $\mathcal{B}$. Then it is easy to show that $S(P_\theta/B)=S(\theta)\cap B$ is a support of P_θ/B for $\mathcal{E}$. Here let us recall that $P_\theta/B(A)=P_\theta(A\cap B)$, $A\in\mathcal{A}$. And it is also a support of $P_\theta/B/\mathcal{B}$ for $\mathcal{E}(\mathcal{B})$. By Lemma 5.5, we have $P_\theta/B\in L(\mathcal{E})$ and $P_\theta/B/\mathcal{B}\in L(\mathcal{E}(\mathcal{B}))$. Let $f(\cdot,\theta)$ be a $\mathcal{B}$-measurable version of the density of P_θ w.r.t. n. Then, since $I_B(\cdot)f(\cdot,\theta)$ is the density of $\{P_\theta/B\}/\mathcal{B}$ w.r.t. $n/\mathcal{B}$, we have $T_n\{P_\theta/B\}/\mathcal{B} = P_\theta/B$. Hence it follows that

$$T_n'\rho(I_A)(\{P_\theta/B\}/\mathcal{B})=\rho(I_A)(T_n\{P_\theta/B\}/\mathcal{B})=\rho(I_A)(P_\theta/B)$$

$$=(T)-\int\rho(I_A)\; dP_\theta/B=P_\theta(A\cap B).$$

On the other hand it holds that

$$T_n'\rho(I_A)(\{P_\theta/B\}/\mathcal{B})=\rho_{\mathcal{B}}(g_A)(\{P_\theta/B\}/\mathcal{B})$$

$$=\;(T)-\int\rho_{\mathcal{B}}(g_A)\; d\{P_\theta/B\}/\mathcal{B}$$

$$=\int g_A \; d\{P_\theta/B\}/\mathcal{B} = \int_B g_A \; dP_\theta/\mathcal{B}.$$

So we have

$$P_\theta(A \cap B) = \int_B g_A \; dP_\theta/\mathcal{B}, \quad A \in \mathcal{A}, B \in \mathcal{B}, \theta \in \Theta.$$

Therefore $\mathcal{B}$ is sufficient for $\mathcal{E}$ and $g_A = E(I_A/\mathcal{B})$ a.e. P_θ for all $\theta \in \Theta$. Here, again by Lemma 5.5, we have $\rho_{\mathcal{B}}(g_A) = \rho_{\mathcal{B}}(E(I_A/\mathcal{B}))$. This completes the proof.

5.4. Application of the extended notion of common conditional probability for PSS subfield

In this section we try to apply the extended notion of common conditional probability for PSS subfield to problems of unbiased estimation in which the common conditional probability for sufficiency plays a central role. Those are Rao-Blackwell theorem, Lehmann-Scheffé theorem etc..

We give in the first place the representation of T_n' for generic point of the M-space of a majorized experiment $\mathcal{E} = (\mathcal{X}, \mathcal{A}, \mathcal{P} = \{P_\theta; \theta \in \Theta\})$.

Lemma 5.6. Let $\mathcal{B}$ be PSS for $\mathcal{E}$. Then for any $f(\mathcal{P}) = \{f(x,\theta); \theta \in \Theta\} \in M(\mathcal{E})$ we have

$$T_n' f(\mathcal{P}) = \{E(f(x,\theta)/\mathcal{B}, P_\theta); \theta \in \Theta\}.$$

Proof. Take any $\theta \in \Theta$. For any $B \in \mathcal{B}$ we have $I_B \cdot d(P_\theta/\mathcal{B})/d(n/\mathcal{B}) \in L^1(\mathcal{X}, \mathcal{B}, n/\mathcal{B})$ and

$$\{I_B \cdot d(P_\theta/\mathcal{B})/d(n/\mathcal{B})\} \cdot n/\mathcal{B} = I_B \cdot (P_\theta/\mathcal{B}) \in L(\mathcal{E}(\mathcal{B})).$$

By Lemma 5.5, $I_B \cdot P_\theta/\mathcal{B}$ belongs to the band generated by $P_\theta/\mathcal{B}$ in the space $\mathfrak{m}(\mathcal{B})$. Then it follows that

$$T_n' f(\mathcal{P})(I_B \cdot P_\theta/\mathcal{B}) = (T)\text{-}\int T_n' f(\mathcal{P}) \; d(I_B \cdot P_\theta/\mathcal{B})$$

$$= \int_B g(x,\theta) \; dP_\theta/\mathcal{B},$$

where $T_n' f(\mathcal{P}) = \{g(x,\theta); \theta \in \Theta\} \in M(\mathcal{E}(\mathcal{B}))$.

On the other hand

$$f(\mathcal{P})\,(T_n I_B \cdot P_\theta/\mathcal{B}) = f(\mathcal{P})\,(I_B\ dP_\theta/dn \cdot n)$$

$$=(T)-\int f(\mathcal{P})\ d(I_B dP_\theta/dn \cdot n)$$

$$= \int f(x,\theta)\ d(I_B dP_\theta/dn \cdot n) = \int_B f(x,\theta)\ dP_\theta$$

$$= \int_B E(f(\cdot,\theta)/\mathcal{B},P_\theta)\ dP_\theta/\mathcal{B}.$$

Hence we have

$$g(x,\theta) = E(f(\cdot,\theta)/\mathcal{B},P_\theta)(x) \text{ a.e. } P_\theta, \text{ for all } \theta\in\Theta.$$

Next theorem is certainly the Rao-Blackwell theorem for PSS subfield.

Theorem 5.10. Let $\mathcal{B}$ be PSS for a majorized experiment $\mathcal{E}$. Then for any $f(\mathcal{P}) = \{f(x,\theta);\theta\in\Theta\}\in M(\mathcal{E})$ there exists a $g(\mathcal{P}) = \{g(x,\theta);\theta\in\Theta\}\in M(\mathcal{E}(\mathcal{B}))$ such that

$$(T)-\int g(\mathcal{P})\ dP_\theta/\mathcal{B} = (T)-\int f(\mathcal{P})\ dP_\theta,$$

$$(T)-\int g^2(\mathcal{P})\ dP_\theta/\mathcal{B} \leqq (T)-\int f^2(\mathcal{P})\ dP_\theta$$

for all $\theta\in\Theta$, where $f^2(\mathcal{P}) \equiv \{f^2(x,\theta);\theta\in\Theta\}$.

Proof. Let us define $g(\mathcal{P})=T_n' f(\mathcal{P}) = \{E(f(\cdot,\theta)/\mathcal{B},P_\theta);\theta\in\Theta\}$.

Then we have, by the definition of (T)-integral, for any $\theta\in\Theta$,

$$(T)-\int g(\mathcal{P})\ dP_\theta/\mathcal{B} = \int g(x,\theta)\ dP_\theta/\mathcal{B} = \int f(x,\theta)\ dP_\theta =$$

$(T)-\int f(\mathcal{P})\ dP_\theta$, and by Jensen's inequality,

$$(T)-\int g^2(\mathcal{P})\ dP_\theta/\mathcal{B} = \int E^2(f(\cdot,\theta)/\mathcal{B},P_\theta)\ dP_\theta/\mathcal{B}$$

$$\leqq \int E(f^2(\cdot,\theta)/\mathcal{B},P_\theta)\ dP_\theta/\mathcal{B} = \int f^2(x,\theta)\ dP_\theta = (T)-\int f^2(\mathcal{P})\ dP_\theta.$$

Remark 5.5. The above proof in particular shows that for

any test function there exists a $\mathcal{B}$-measurable "test function" which has the same "power".

Remark 5.6. In the above proof we only used the (T)-integral w.r.t. P_θ (or $P_\theta/\mathcal{B}$). Hence the proof is completely the same as the usual proof of the Rao-Blackwell theorem.

Remark 5.7. Even if $f(\mathcal{P}) = \{f;\theta\in\Theta\}$, by Theorem 5.9, $g(\mathcal{P})$ may not have the same form $\{g;\theta\in\Theta\}$ if $\mathcal{B}$ is not sufficient.

Remark 5.8. Let us recall the Rao-Blackwell theorem in Theorem 4.4. The result was heavily indebted to Lemma 2.3. There we considered the problem how largely we should expand the given PSS subfield to get a sufficient subfield. And we applied the common conditional probability for the expanded sufficient subfield to get a Rao-Blackwell theorem for the original PSS subfield. But we confirmed in Remark 4.3 that Lemma 2.3 and the "only if" part of Theorem 4.4 themselves do not necessarily hold true in a majorized experiment. On the other hand it is easy to show the "only if" part of Theorem 4. 4 using Theorem 5.10. Hence Theorem 5.10 can be interpreted as a correspondent of the usual Rao-Blackwell theorem for sufficiency. But in this case we must restrict to majorized experiments and bounded "statistics". Anyway it seems to exist a layer for constructing a theory of sufficiency or PSS in undominated cases at this stage of the study.

Next definition gives an extended notion of the boundedly complete statistic. This definition is a rewriting of the definition due to Siebert(1979a).

Definition 5.5. Let $\mathcal{E}=(\mathcal{X},\mathcal{A},\mathcal{P}=\{P_\theta;\theta\in\Theta\})$ be any experiment and $\mathcal{B}$ be any subfield of $\mathcal{A}$. If the following is correct then $\mathcal{B}$ is said to be *strongly boundedly complete*; for any $f(\mathcal{P})\in M(\mathcal{E}(\mathcal{B}))$ such that

$$(T)-\int f(\mathcal{P})\ dP_\theta/\mathcal{B} = 0,\ \theta\in\Theta, \tag{5.10}$$

it follows that

$$(T)-\int f(\mathcal{P})\ d\mu = 0 \text{ for all } \mu\in L(\mathcal{E}(\mathcal{B})).$$

<u>Remark 5.9</u>. $\mathcal{B}$ is strongly boundedly complete if and only if $f(x,\theta) = 0$ a.e. P_θ for all $\theta\in\Theta$ hold for any $f(\mathcal{P}) = \{f(x,\theta);\ \theta\in\Theta\}\in \mathcal{M}(\mathcal{E}(\mathcal{B}))$ satisfying (5.10). This follows from the fact that $f(x,\theta)\cdot P_\theta/\mathcal{B}$ belongs to the band generated by $\{P_\theta/\mathcal{B}\}$.

<u>Remark 5.10</u>. Let $\mathcal{B}$ be strongly boundedly complete and f be a bounded $\mathcal{B}$-measurable function. Suppose that $\int f(x)\ dP_\theta/\mathcal{B} = 0$ for all $\theta\in\Theta$. Then applying Remark 5.9 to $\{f;\ \theta\in\Theta\}$ we have $f(x) = 0$ a.e. P_θ for all $\theta\in\Theta$. This is the usual definition of bounded completeness of $\mathcal{B}$. For an example of subfield which is not strongly boundedly complete, but boundedly complete we refer to Torgersen(1991, Section 7.3); it is the experiment in which the sample space is the standard Borel space of the unit interval, and the family of probability measures consists of all the point probability measures and the uniform distribution on the unit interval. This experiment is not majorized. Then $\mathcal{A}$ is boundedly complete, but not strongly boundedly complete. An example in the case of majorized experiment does not seem to be known. It can be shown that if $\mathcal{B}$ is PSS and $\mathcal{E}(\mathcal{B})$ is weakly dominated, then $\mathcal{B}$ is boundedly complete if and only if $\mathcal{B}$ is strongly boundedly complete.

<u>Remark 5.11</u>. Here is a mathematical characterization of strongly bounded completeness. The following conditions are all equivalent.

(1) $\mathcal{B}$ is strongly boundedly complete.

(2) $\mathcal{P}/\mathcal{B}$ is total in $\mathcal{M}(\mathcal{E}(\mathcal{B}))$.

(3) $T(\mathcal{M}(\mathcal{E}(\mathcal{B})),\mathcal{P}/\mathcal{B})$ is a T_1-space.

This follows, for example, from Theorem 3.81-B of Taylor (1958).

Next theorem is a variant of a theorem of Lehmann and Scheffé which shows existence of uniformly minimum variance unbiased estimator based upon a complete sufficient statistic, and the proof of which is again essentially the same as the proof of Lehmann and Scheffé's theorem for sufficiency.

<u>Theorem 5.11</u>. Let $\mathcal{B}$ be PSS for a majorized experiment $\mathcal{E}=(\mathcal{X},\mathcal{A},\mathcal{P}=\{P_\theta;\theta\in\Theta\})$. Let $\mathcal{B}$ be strongly boundedly complete. Then for any $f(\mathcal{P})=\{f(x,\theta);\theta\in\Theta\}\in\mathcal{M}(\mathcal{E})$ there exists a $g(\mathcal{P})= \{g(x,\theta);\ \theta\in\Theta\}\in\mathcal{M}(\mathcal{E}(\mathcal{B}))$ such that

(a) $(T)-\int g(P)\ dP_\theta/\mathcal{B} = (T)-\int f(P)\ dP_\theta$ for all $\theta\in\Theta$,

(b) for any $h(P) = \{h(x,\theta);\theta\in\Theta\}\in\mathcal{M}(\mathfrak{X})$ such that

$$(T)-\int h(P)\ dP_\theta = (T)-\int f(P)\ dP_\theta \text{ for all } \theta\in\Theta,$$

it holds that

$$(T)-\int g^2(P)\ dP_\theta \le (T)-\int h^2(P)\ dP_\theta \text{ for all } \theta\in\Theta.$$

Proof. Let us define $g(P) = T_n' f(P) = \{E(f(\cdot,\theta)/\mathcal{B},P_\theta);\ \theta\in\Theta\}$. Then we have

$$(T)-\int g(P)\ dP_\theta/\mathcal{B} = \int E(f(\cdot,\theta)/\mathcal{B},P_\theta)\ dP_\theta/\mathcal{B} = \int f(x,\theta)\ dP_\theta$$

$$= (T)-\int f(P)\ dP_\theta,$$

and

$$(T)-\int f(P)\ dP_\theta = (T)-\int h(P)\ dP_\theta$$

$$= \int h(x,\theta)\ dP_\theta = \int E(h(\cdot,\theta)/\mathcal{B},P_\theta)\ dP_\theta/\mathcal{B}.$$

By strongly boundedly completeness of $\mathcal{B}$ we have

$$g(x,\theta) = E(h(\cdot,\theta)/\mathcal{B},P_\theta) \text{ a.e. } P_\theta \text{ for all } \theta\in\Theta.$$

By Jensen's inequality,

$$g^2(x,\theta) \le E(h^2(\cdot,\theta)/\mathcal{B},P_\theta) \text{ a.e. } P_\theta.$$

So it follows that

$$(T)-\int g^2(P)\ dP_\theta/\mathcal{B} = \int g^2(x,\theta)\ dP_\theta/\mathcal{B}$$

$$\le \int E(h^2(\cdot,\theta)/\mathcal{B},P_\theta)\ dP_\theta/\mathcal{B}$$

$$= \int h^2(x,\theta)\ dP_\theta/\mathcal{B} = (T)-\int h^2(P)\ dP_\theta \text{ for all } \theta\in\Theta.$$

Let us recall the following theorem which is also due to

Lehmann and Scheffé : If $\mathcal{B}$ is sufficient and boundedly complete then $\mathcal{B}$ is the smallest sufficient subfield. When we consider a PSS instead of a sufficient subfield we are forced to change the definition of "the smallest".

<u>Definition 5.6</u>. A PSS subfield $\mathcal{B}$ for a majorized experiment $\mathcal{E}=(\mathcal{X},\mathcal{A},\mathcal{P}=\{P_\theta;\theta\in\Theta\})$ is said to be *pairwise smallest* if, for any other PSS $\mathcal{C}$, it holds that $\mathcal{B}\subset\tilde{\mathcal{C}}(\mathcal{A})$.

<u>Theorem 5.12</u>. If $\mathcal{B}$ is PSS for a majorized experiment $\mathcal{E}=(\mathcal{X},\mathcal{A},\mathcal{P}=\{P_\theta;\theta\in\Theta\})$ and strongly boundedly complete then it is the pairwise smallest PSS subfield.

<u>Proof</u>. Let $\mathcal{C}$ be any PSS for $\mathcal{E}$ and define

$$T_n(\mathcal{C}):\ L(\mathcal{E}(\mathcal{C}))\to L(\mathcal{E})\ \text{by}$$
$$f\cdot n/\mathcal{C}\to f\cdot n,\quad f\in L^1(\mathcal{X},\mathcal{C},n/\mathcal{C}),$$

where n is any pivotal measure for $\mathcal{E}$. Let $T_n'(\mathcal{C})$ be its dual mapping. $T_n(\mathcal{B})$ and $T_n'(\mathcal{B})$ are defined in the same way as $T_n(\mathcal{C})$ and $T_n'(\mathcal{C})$ respectively. Since $M(\mathcal{E}(\mathcal{C})), M(\mathcal{E}(\mathcal{B}))\subset M(\mathcal{E})$, for any $B\in\mathcal{B}$, $\{I_B;\theta\in\Theta\}\in M(\mathcal{E})$ and $T_n'(\mathcal{B})(T_n'(\mathcal{C})(\{I_B;\theta\in\Theta\}))\in M(\mathcal{E}(\mathcal{B}))$. Then it follows that

$$(T)-\int T_n'(\mathcal{B})\ (T_n'(\mathcal{C})(\{I_B;\theta\in\Theta\}))\ dP_\theta/\mathcal{B}$$
$$=\int E(E(I_B/\mathcal{C},P_\theta)/\mathcal{B},P_\theta)\ dP_\theta/\mathcal{B}$$
$$=\int E(I_B/\mathcal{C},P_\theta)\ dP_\theta=\int I_B\ dP_\theta$$
$$=(T)-\int T_n'(\mathcal{B})(\{I_B;\theta\in\Theta\})\ dP_\theta/\mathcal{B},$$

so that

$$(T)-\int [T_n'(\mathcal{B})\{T_n'(\mathcal{C})(\{I_B;\theta\in\Theta\})\}$$
$$-\ T_n'(\mathcal{B})(\{I_B;\theta\in\Theta\})]\ dP_\theta/\mathcal{B}=0$$

for all $\theta\in\Theta$. Since, by the assumption, $\mathcal{B}$ is strongly boundedly complete,

$$T_n'(\mathcal{B})\{T_n'(\mathcal{C})(\{I_B;\theta\in\Theta\})\} - T_n'(\mathcal{B})(\{I_B;\theta\in\Theta\})$$

$$= \{E(E(I_B/\mathcal{C},P_\theta)/\mathcal{B},P_\theta) - E(I_B/\mathcal{B},P_\theta);\theta\in\Theta\}$$

$$= \{E(\{E(I_B/\mathcal{C},P_\theta) - I_B\}/\mathcal{B},P_\theta);\theta\in\Theta\} = 0$$

as an element of $\mathcal{M}(\mathcal{E}(\mathcal{B}))$. Then

$$E(\{E(I_B/\mathcal{C},P_\theta) - I_B\}/\mathcal{B},P_\theta) = 0 \text{ a.e. } P_\theta/\mathcal{B} \text{ for all } \theta\in\Theta.$$

So we have

$$\int_B I_B dP_\theta/\mathcal{B} = \int I_B E(E(I_B/\mathcal{C},P_\theta)\, dP_\theta/\mathcal{B}$$

$$= \int E(I_B E(I_B/\mathcal{C},P_\theta))\, dP_\theta/\mathcal{B} = \int I_B E(I_B/\mathcal{C},P_\theta)\, dP_\theta.$$

This implies

$$\int I_B(1 - E(I_B/\mathcal{C},P_\theta))\, dP_\theta = 0. \qquad (5.11)$$

On the other hand

$$\int I_B\, dP_\theta = \int E(I_B/\mathcal{C},P_\theta)\, dP_\theta/\mathcal{C}.$$

Therefore it follows that

$$\int (1 - I_B)E(I_B/\mathcal{C},P_\theta)\, dP_\theta = 0. \qquad (5.12)$$

From (5.11) and (5.12),

$$I_B(1 - E(I_B/\mathcal{C},P_\theta)) = 0 \text{ a.e. } P_\theta, \qquad (5.13)$$

$$(1 - I_B)E(I_B/\mathcal{C},P_\theta) = 0 \text{ a.e. } P_\theta. \qquad (5.14)$$

By (5.13) and (5.14),

$$I_B = E(I_B/\mathcal{C},P_\theta) \text{ a.e. } P_\theta \text{ for all } \theta\in\Theta. \qquad (5.15)$$

This implies that $B\in\hat{\mathcal{C}}(\mathcal{A})$. Since $\mathcal{C}$ is pairwise sufficient $\tilde{\mathcal{C}}(\mathcal{A})$

$=\hat{\mathscr{C}}(\mathcal{A})$ as was noted in the paragraph before Corollary 4.2. Hence we have $B\in\tilde{\mathscr{C}}(\mathcal{A})$, which implies $\mathcal{B}\subset\tilde{\mathscr{C}}(\mathcal{A})$. This completes the proof.

The above proof, though the notions of L- and M-space are used, is essentially based upon the method of Bahadur (1957).

It is clear that, if $\mathcal{B}$ is the smallest PSS, then it is the pairwise smallest PSS. But, as is shown in the next example, there exists subfield $\mathcal{B}$ which is strongly boundedly complete and is the pairwise smallest PSS, but is not the smallest PSS. This shows that in Theorem 5.12 we can not replace the pairwise smallest PSS with the smallest PSS.

Example 5.2. Let $\mathcal{X}=R^1$ and $\mathcal{A}=2^{\mathcal{X}}$. Let $\mathcal{P}$ be the totality of point probability measures P_θ, $\theta\in\Theta=\mathcal{X}$. This experiment $\mathscr{E}$ is discrete. By Theorem 3.7, the subfield $\{A\subset\mathcal{X};\ A$ is countable or $\mathcal{X}-A$ is countable$\}$ is the smallest PSS. Let $\mathcal{B}$ be the Borel field. Hence $\mathcal{B}$ is PSS and is not the smallest PSS. We show that $\mathcal{B}$ is strongly boundedly complete. Therefore, by Theorem 5.12, $\mathcal{B}$ is the pairwise smallest PSS. Let $f(\mathcal{P})=\{f(x,\theta);\theta\in\Theta\}\in M(\mathscr{E}(\mathcal{B}))$ and let $(T)-\int f(\mathcal{P})\ dP_\theta/\mathcal{B} = 0$ for all $\theta\in\Theta$. Then we have $f(\theta,\theta) = 0$ for all $\theta\in\Theta$. Take any positive $\mu\in L(\mathscr{E}(\mathcal{B}))$. Then there exists a countable subfamily $\{P_{\theta_1},P_{\theta_2},\ldots\}$ of $\mathcal{P}$ such that μ belongs to the band generated by $\{P_{\theta_1}/\mathcal{B},P_{\theta_2}/\mathcal{B},\ldots\}$ in $\mathfrak{m}(\mathcal{B})$. So, for this countable subfamily, there exists a $\mathcal{B}$-measurable function f such that $f(x) = f(x,\theta_i)$ a.e. P_{θ_i} for all $i\geq1$, which implies

$$f(\theta_i) = f(\theta_i,\theta_i) = 0 \text{ for all } i\geq1. \tag{5.16}$$

Since $\mu\in L((\mathcal{X},\mathcal{B},\{P_{\theta_i}/\mathcal{B};i\geq1\}))$ it follows that $\mu\ll\{P_{\theta_i}/\mathcal{B};i\geq1\}$ by Lemma 5.5. Let us define $B = \mathcal{X} - \{\theta_1,\theta_2,\ldots\}$. Then we have $P_{\theta_i}(B) = 0$ for all $i\geq1$, therefore $\mu(B)=0$. Hence it follows that, by (5.16),

$$(T)-\int f(\mathcal{P})\ d\mu = \int f(x)\ d\mu = \sum_{i=1}^{\infty} f(\theta_i)\mu(\theta_i) = 0.$$

For generic point of $L(\mathscr{E}(\mathcal{B}))$ we have the same result by Jordan decomposition of a signed measure. Hence $f(\mathcal{P}) = 0$ as an element of $M(\mathfrak{E}(\mathcal{B}))$, which implies that $\mathcal{B}$ is strongly boundedly complete.

5.5. Bibliographical notes

The notion of deficiency was introduced by Le Cam (1964). For general theory of comparison of experiments we refer to Siebert(1979b), Heyer(1982), Strasser(1985), Le Cam(1986), Bomze(1990), Mussmann(1991), and Torgersen(1976,1991). Schaefer(1974), Fremlin(1974), and Torgersen(1991) contain many results on L- and M-spaces. Theorem 5.2 is due to Siebert (1979a). Theorem 5.3 and its converse are due to Littaye, Petit,Piednoir and van Cutsem (1969). The present proof was communicated to the author by Luschgy. We also refer to Luschgy(1978), Heyer and Kusama(1991),and Torgersen(1991, Proposition 6.2.28). For a PSS subfield the transition T_n which makes the deficiency zero was found by Yamada(1985).

Representation of the dual space of a general L^1-space is due to McShane(1962). Lemma 5.5 is a technically important result and was proved by Luschgy and Mussmann(1985). A more detailed discussion on $U(\mathfrak{E})$ and $V(\mathfrak{E})$ can be found in Ramamoorthi and Roy(1987), and Luschgy(1988). Torgersen(1979) showed a representation theorem for the M-space of any experiment by the (T)-integral. Theorem 5.5 and Section 5.3 were quoted from Yamada(1988). Yamada(1991) applied the common conditional probability for PSS subfield to the Rao-Blackwell, Lehmann-Scheffé theorem etc. as was presented in Section 5.4.

Blackwell(1951,1953) founded the theory of comaprison of experiments based on the definition, due to Bohnenblust, Shapley and Sherman(1949), that an experiment $\mathfrak{E}$ is more informative than another experiment $\mathfrak{F}$ if both of them have the same parameter space and $\mathfrak{F}$ is obtained by a stochatic transformation from $\mathfrak{E}$. We refer to the same notion of "exhaustive" in the Bibliographical notes in Chapter 4. Le Cam (1964) extended this notion to the notion of deficiency (Definition 5.1(4)) and succeeded in showing especially the equivalency between the comparison based on decision problems and the comparison based on sufficiency.

A sufficient sublattice of a experiment $\mathfrak{E}$, which was introduced by Le Cam(1964), is defined as a weakly closed vector sublattice H of $M(\mathfrak{E})$ containing the unit u onto which there exists a positive linear projection Π from $M(\mathfrak{E})$ such that

$$\Pi e = e \text{ and } \langle \Pi u, P_\theta\rangle = \langle u, P_\theta\rangle \text{ for all } u \in M(\mathscr{E}) \text{ and } \theta \in \Theta.$$

Then it was proved by Le Cam(1964,Proposition 9) that

> *Π has the characteristic property of the conditional expectation that $\Pi(uv) = v\Pi(u)$ for every $u \in M(\mathscr{E})$ and $v \in H$.*

Moreover there exists always the smallest sufficient sublattice, and a sublattice is sufficient if and only if it is "pairwise sufficient". Thus, as was quoted in Introduction, Le Cam(1964) succeeded in overcoming the Burkholder and Pitcher problems.

Relationship between the notion of the sufficient sublattice and pairwise sufficiency was studied by Littaye ,Petit,Piednoir and van Cutsem(1969), and Siebert(1979a). Then $\mathscr{B}$ is pairwise sufficient if and only if the weak closure of the standard embedding of the bounded $\mathscr{B}$-measurable functions is a sufficient sublattice of $\mathscr{E}$. However, to make the situation complicated, there exist sufficient sublattices which do not correspond to a pairwise sufficiency.

The method of the construction of the "common conditional probability" by Le Cam(1964,Proposition 9) is based on the smoothing technique applied to the dual D of a transition which renders the deficiency zero. Then the sufficiency projection, i.e. the "common conditional probability" Π thus obtained satisfies

$$D\Pi = \Pi D = \Pi \tag{5.17}$$

Such a technique for a construction of the "common conditional probabilty " as Le Cam had considered was studied by Pfanzagl and Weber (1973), Roy and Ramamoorthi (1979). Pfanzagl and Weber (1973) proved the existence of Π in (5.17) in which D is homogeneous, translation invariant, monotone and expectation invariant operator. Roy and Ramamoorthi (1979) applied the same technique of averaging to show that any D-sufficient subfield contains a sufficient subfield. For further discussions we refer to Kudo(1972). The notion of statistical operation, including the "common conditional probability", was studied, from a very broad point of view, by

Dawid (1980).

Le Cam(1964, Theorem 3 and Proposition 7), to be more tremendous, asserts the existence of "approximate common conditional probability" for an "approximate sufficient" (ε-deficient) experiment for another experiment with the same parameter space as the former. We also refer to Strasser(1985, Theorem 45.6). The proof is based on a general minimax theorem (Le Cam (1964, Theorem 2), Strasser (1985, Theorem 45.6)).

Heyer and Yamada(1993) considered a notion of the "common conditional probability" for pairwise sufficient subfields. Let $\mathscr{E}=(\mathscr{X},\mathscr{A},\mathscr{P}=\{P_\theta;\theta\in\Theta\})$ be a majorized experiment and let $\mathscr{B}$ be pairwise sufficient. Then $f_A(\mathscr{P})=\{E(I_A/\mathscr{B},P_\theta);\theta\in\Theta\}$, for any $A\in\mathscr{A}$, is a bounded countably coherent family of $\mathscr{B}$-measurable functions. Hence $f_A(\mathscr{P})\in\mathscr{M}(\mathscr{E}(\mathscr{B}))$. Let's define a mapping

$T:L(\mathscr{E}(\mathscr{B}))\longrightarrow L(\mathscr{E})$ by

$$(T\mu)(A)=\langle f_A(\mathscr{P}),\mu\rangle\ ,\quad \mu\in L(\mathscr{E}(\mathscr{B}))\,,\ A\in\mathscr{A}.$$

Then T is a transition which makes the deficiency of $\mathscr{E}(\mathscr{B})$ w.r.t. $\mathscr{E}$ zero. T is an extended notion of T_n in Section 5.3 in the sense that $T=T_n$ if $\mathscr{B}$ is PSS. By taking dual of T, we have the same result as Theorem 5.8 except (1). This T is not only a transition which renders the deficiency zero, but also it has the characteristic property of the conditional probability that

$$TI_BP_\theta/\mathscr{B} = I_BTP_\theta/\mathscr{B} \tag{5.18}$$

for all $\theta\in\Theta$ and $B\in\mathscr{B}$. Hence the dual T' of T is idempotent, i.e. $T'^2=T'$. By this property of T' we need not take the smoothing technique. Namely $T'=\Pi$. Conversely if $\Pi=T'$ then T satisfies (5.18).

Luschgy(1988) characterized Le Cam's sufficient sublattice as the space of all fixed points for some subsemigroup of the totality of positive, linear, constant preserving, and expectation invariant operators on $\mathscr{M}(\mathscr{E})$. In particular for any pairwise sufficient subfield the pairwise completion coincides with some (extended notion of) invariant subfield, extending a result of Trenkler (1982).

Kusama(1991) studied class properties of statistical experiments under an equivalency.

Chapter 6. Structure of pivotal measure

The aim of this chapter is to investigate the structure of pivotal measures using the L-space of the experiment. We show a pivotal measure is a measure which is represented as the sum of elements of a maximal orthogonal system of the L-space of the marginal experiment corresponding to the smallest PSS subfield. This result covers our previous construction of pivotal measures in majorized experiments (Section 2.1), and the classical Halmos and Savage's pivotal measure $\sum_{i=1}^{\infty} 2^{-i} P_{\theta_i}$ in dominated experiment (Section 2.4).

The main reference of this chapter is Luschgy, Mussmann and Yamada(1988).

6.1. Minimal L-space

In this section we introduce some preliminary results from the L-space theory. Let $\mathcal{E}=(\mathcal{X},\mathcal{A},\mathcal{P}=\{P_\theta;\theta\in\Theta\})$ be a majorized experiment. Let $L_m(\mathcal{E})$, *minimal L-space* of $\mathcal{E}$, be the closed vector sublattice of $\mathfrak{M}(\mathcal{A})$ generated by $\mathcal{P}$. Note that $L_m(\mathcal{E})$ is itself an L-space.

Suppose that $\mathcal{P}^*$ is the totality of probability measures in $L_m(\mathcal{E})$ and let $\mathcal{E}^*=(\mathcal{X},\mathcal{A},\mathcal{P}^*)$. For a subfield $\mathcal{B}$, $W(\mathcal{B})$ denotes the set of all $w\in\mathfrak{M}(\mathcal{A})$ of the form

$$w=f\cdot(\sum_{i=1}^{\infty} 2^{-i} P_{\theta_i}),$$

where f is a $\mathcal{B}$-measurable function and $P_{\theta_i}\in\mathcal{P}$ for all $i\geq 1$.

Proposition 6.1. Let $\mathcal{B}$ be a subfield of $\mathcal{A}$.
(a) If each $P_\theta\in\mathcal{P}$ has a support for $\mathcal{E}$ belonging to $\mathcal{B}$, then each $w\in W_+(\mathcal{B})$ has also a support for $\mathcal{E}$ which belongs to $\mathcal{B}$. Here $W_+(\mathcal{B}) = \{w\in W(\mathcal{B});\ w>0\}$.
(b) If $\mathcal{B}$ is pairwise sufficient for $\mathcal{E}$, then $W(\mathcal{B})$ is a closed vector sublattice of $\mathfrak{M}(\mathcal{A})$ and $\mathcal{B}$ is pairwise sufficient for the set of all probability measures from $W(\mathcal{B})$.

Proof. (a) Let us define

$$w=f\cdot(\sum_{i=1}^{\infty}2^{-i}P_{\theta_i})\in W_+(\mathcal{B}),$$

where f is $\mathcal{B}$-measurable. Let T_i be a support of P_{θ_i} for $\mathcal{E}$ belonging to $\mathcal{B}$ for each $i\geq 1$. Then $[f(x)>0]\cap(\bigcup_{i=1}^{\infty}T_i)$ is a support of w for $\mathcal{E}$. It is clear that $w(\mathcal{X}-[f(x)>0]\cap(\bigcup_{i=1}^{\infty}T_i))=0$. Suppose that $A\subset[f(x)>0]\cap(\bigcup_{i=1}^{\infty}T_i)$ and $w(A)=0$. Then by

$$0=w(A)=\int_A f\ d(\sum_{i=1}^{\infty}2^{-i}P_{\theta_i})=\sum_{i=1}^{\infty}2^{-i}\int_{A\cap T_i}f\ dP_{\theta_i},$$

it follows that $P_{\theta_i}([f(x)>0]\cap A\cap T_i)=0$ for all $i\geq 1$. Since T_i is a support of P_{θ_i} for $\mathcal{E}$ we have $P_\theta([f(x)>0]\cap A\cap T_i)=0$ for all $\theta\in\Theta$. By $A=\bigcup_{i=1}^{\infty}([f(x)>0]\cap A\cap T_i)$, we have $P_\theta(A)=0$.

(b) $\mathcal{B}$ is sufficient for any dominated subset of $\mathcal{P}$, because it is pairwise sufficient for $\mathcal{P}$. Then by the discussion on pivotal measure in Section 2.4, we can assume that $dP_\theta/d(\sum_{i=1}^{\infty}2^{-i}P_{\theta_i})$ is $\mathcal{B}$-measurable whenever $P_\theta(\in\mathcal{P})$ is absolutely continuous w.r.t. $\sum_{i=1}^{\infty}2^{-i}P_{\theta_i}$. From this we obtain that, for each sequence $\{w_j\}$ from $W_+(\mathcal{B})$, there exists a sequence $\{f_j\}$ of $\mathcal{B}$-measurable functions and a sequence $\{P_{\theta_i}\}$ in $\mathcal{P}$ such that

$$w_j=f_j\cdot(\sum_{i=1}^{\infty}2^{-i}P_{\theta_i})\quad\text{for all } j\geq 1.$$

From these representations it can be shown that $W(\mathcal{B})$ is a closed vector sublattice of $\mathfrak{m}(\mathcal{A})$. To show that $\mathcal{B}$ is pairwise sufficient for the set of all probability measures from $W(\mathcal{B})$, it is enough to prove that $\mathcal{B}$ is sufficient for any set of the form of probability measures $w=f\cdot(\sum_{i=1}^{\infty}2^{-i}P_{\theta_i})$, where f is $\mathcal{B}$-measurable and the sequence $\{P_{\theta_i}\}$ is fixed. For any such $w\in W(\mathcal{B})$ we get

$$\int_B I_A\ dw=\int_B I_A f\ d(\sum_{i=1}^{\infty}2^{-i}P_{\theta_i})=\int_B E(I_A/\mathcal{B})f\ d(\sum_{i=1}^{\infty}2^{-i}P_{\theta_i})$$

$$= \int_B E(I_A/\mathcal{B})\ dw$$

for all $A\in\mathcal{A}$ and $B\in\mathcal{B}$, where $E(I_A/\mathcal{B})$ is the common conditional probability of A given $\mathcal{B}$ for the sequence $\{P_{\theta_i}\}$, that is,

$$E(I_A/\mathcal{B})(x) = E(I_A/\mathcal{B},P_{\theta_i})(x) \text{ a.e. } P_{\theta_i} \text{ for all } i\geq 1.$$

<u>Corollary 6.1</u>. Suppose that $\mathcal{B}$ is pairwise sufficient for $\mathcal{E}$. Then we have:
(a) If each $P\in\mathcal{P}$ has a support for $\mathcal{E}$ belonging to $\mathcal{B}$, then each $Q\in\mathcal{P}^*$ has also a support for $\mathcal{E}$ belonging to $\mathcal{B}$.
(b) $\mathcal{B}$ is pairwise sufficient for $\mathcal{E}^*$.

To show these it is enough to note that $W(\mathcal{B})\supset L_m(\mathcal{E})$, because $W(\mathcal{B})$ is a closed vector sublattice of $\mathfrak{M}(\mathcal{A})$ including $\mathcal{P}$ by Proposition 6.1 (b). Note that the converse of (a) is also true because $\mathcal{P}^*\supset\mathcal{P}$.

Suppose that $\mathcal{H}$ is a family of $\mathcal{A}$-measurable functions. Then $(\mathcal{H})$ denotes the smallest subfield of $\mathcal{A}$ which makes each $h\in\mathcal{H}$ measurable. For two families of sets, $\mathcal{S}_1$ and $\mathcal{S}_2$, $\mathcal{S}_1\vee\mathcal{S}_2$ denotes the smallest σ-field including both $\mathcal{S}_1$ and $\mathcal{S}_2$.

<u>Lemma 6.1</u>. Suppose that m is a majorizing measure for a majorized experiment $\mathcal{E}=(\mathcal{X},\mathcal{A},\mathcal{P}=\{P_\theta;\theta\in\Theta\})$. Then we have

$$\sigma(dQ/dm;Q\in\mathcal{P}^*)\vee N(\mathcal{P}) = \sigma(dP_\theta/dm;\theta\in\Theta)\vee N(\mathcal{P}). \tag{6.1}$$

<u>Proof</u>. In the first place we note that, since $\mathcal{E}$ is majorized by m, $\mathcal{E}^*$ is also majorized by m. To show this note that $L(m)=\{f\cdot m;f\in L^1(\mathcal{X},\mathcal{A},m)\}$ is a band including $\mathcal{P}$, and hence is a closed vector sublattice including $\mathcal{P}$. So we have $L_m(\mathcal{E})\subset L(m)$. Hence $\mathcal{E}^*$ is majorized by m. Then it is also clear that the left hand side of (6.1) includes the right hand side. On the other hand $\{f\cdot m;f\in L^1(\mathcal{X},\sigma(dP_\theta/dm;\theta\in\Theta)\vee N(\mathcal{P}),m)\}$ is a closed vector sublattice of $L(m)$ and includes $\mathcal{P}$. Hence it includes $L_m(\mathcal{E})$. This shows the converse inclusion is also true.

We quote a result from the L-space theory to look for the densities for the candidate of a pivotal measure, that is, for the sum of elements of a maximal orthogonal system of $L_m(\mathscr{E})$.

Suppose that V is an L-space. We define

$$\Pi_x(y) = sup(y\wedge kx;k>0)$$

for all $x,y\in V_+$. This $\Pi_x(y)$ is the projection of y onto the band generated by $\{x\}$. Furthermore we note that every L-space V has a maximal orthogonal system in V_+, which can be shown by Zorn's lemma.

Lemma 6.2. Suppose that V is an L-space.
(a) If $\{x_i;i\in I\}$ is an increasing net in V_+ such that $sup\{\|x_i\|;i\in I\}<\infty$, then $lim_i x_i$ exists and is equal to $sup_i x_i$.
(b) If D is a maximal orthogonal system in V and if $y\in V_+$, then

$$y=sup\{\Sigma\{\Pi_u(y);u\in\Lambda\};\Lambda\}=lim\{\Sigma\{\Pi_u(y);u\in\Lambda\};\Lambda\},$$

where Λ ranges through all finite subsets of D. The set $\{u\in D;\Pi_u(y)\neq 0\}$ is countable for each such y.

Proof. (a) That $sup\{x_i;i\in I\}$ exists is the result of order completeness property of an L-space. Then it is easy to show this $sup\{x_i;i\in I\}$ is the limit of $\{x_i;i\in I\}$.
(b) By Lemma 3.5 of Kakutani (1941), we have

$$\Sigma\{\Pi_u(y);u\in\Lambda\} = \Pi_{\Sigma\{u;u\in\Lambda\}}(y)\leq y.$$

Then, by (a), it holds that

$$sup\{\Sigma\{\Pi_u(y);u\in\Lambda\};\Lambda\} = lim\{\Sigma\{\Pi_u(y);u\in\Lambda\};\Lambda\}.$$

By Riesz's decomposition theorem (Schaefer (1974), Chapter 2, 2.10), the band in V generated by D is equal to V. From Proposition 2.1 in Chapter 2 of Schaefer(1974) we have

$$sup\{\Pi_{\Sigma\{u;u\in\Lambda\}}(y);\Lambda\} = y.$$

If $\Pi_u(y) \neq 0$ for all u from an uncountable subset of D, there exists an $\varepsilon>0$ such that $\{u\in D;\|\Pi_u(y)\|\geq\varepsilon\}$ is an infinite set. Because $\|y\| \geq \Sigma\{\|\Pi_u(y)\|;u\in\Lambda\}$ for all Λ, this is a contradiction.

If $u, v \in L_m(\mathcal{E})_+$, then $\Pi_v(u) \in L_m(\mathcal{E})$ by Lemma 6.2(a). Then we have

$$\Pi_v(u) = (du/dm \cdot I_{[dv/dm>0]}) \cdot m, \text{ where } m=u+v.$$

This can be shown using the vector lattice isomorphism between $L^1(\mathcal{X},\mathcal{A},m)$ and $L(m)$.

Lemma 6.3. If D is a maximal orthogonal system of $L_m(\mathcal{E})$, then it is also a maximal orthogonal system of $L(\mathcal{E})$.

Proof. D is an orthogonal system of $L(\mathcal{E})$ because $L_m(\mathcal{E})$ is a sublattice of $L(\mathcal{E})$. Let $B(D)$ be the band generated by D in $L(\mathcal{E})$. Then $B(D) \cap L_m(\mathcal{E})$ is a band in $L_m(\mathcal{E})$ and includes D. Hence we have $B(D) \cap L_m(\mathcal{E}) = L_m(\mathcal{E})$, which implies $B(D) \supset L_m(\mathcal{E})$. Then it follows that $B(D) \supset L_m(\mathcal{E}) \supset \mathcal{P}$. So we have $L(\mathcal{E}) = B(D)$. This shows that D is a maximal orthogonal system of $L(\mathcal{E})$.

Let us define a measure ν_D on $(\mathcal{X},\mathcal{A})$, which is the sum of elements of D, by

$$\nu_D(A) = sup \{\Sigma\{w(A); w \in F\};\ F \subset D, \text{ and } F \text{ is finite}\},\ A \in \mathcal{A}.$$

By Lemma 6.3 and Lemma 5.5, ν_D is an equivalent majorizing measure for $\mathcal{E}$.

Proposition 6.2. Suppose that D is a maximal orthogonal system of $L_m(\mathcal{E})$. Then for each $v \in L_m(\mathcal{E})_+$ there exists a countable subset $D_0 \subset D$ such that

$$\Sigma\{I_{S(w)} d(\Pi_w(v))/dw; w \in D_0\}$$

is a density of v w.r.t. ν_D. Here $S(w)$ is a support of w.

Proof. By Lemma 6.2(b) there is a countable subset $D_0 \subset D$ such that $v = \Sigma\{\Pi_w(v); w \in D_0\}$. Then we have

$$dv/d\nu_D = \Sigma\{d(\Pi_w(v))/d\nu_D; w \in D_0\} = \Sigma\{d\Pi_w(v)/dw \cdot I_{S(w)}; w \in D_0\},$$

because $\Pi_w(v)$ is concentrated on $S(w)$, and $\nu_D = w$ on $S(w)$.

Example 6.1. (a) Suppose that $\mathcal{E}$ is dominated. Then there is a dominating measure of the form $\nu=\sum_{i=1}^{\infty} 2^{-i}P_{\theta_i}$, $P_{\theta_i}\in\mathcal{P}$, by Corollary 2.1. Then $D=\{\nu\}$ is a maximal orthogonal system of $L_m(\mathcal{E})$. To show this suppose that there is an orthogonal system D_1 such that $D_1\supsetneqq D$. Take any $\nu_1\in D_1-D$. Then there exists a set $A\in\mathcal{A}$ such that $\nu_1(\mathcal{X}-A)=\nu(A)=0$. Hence $A\in N(\mathcal{P})$. But $\nu_1\ll\mathcal{P}$ by Lemma 5.5, which implies $\nu_1(A)=0$. This is a contradiction.
(b) Let $\mathcal{E}=(\mathcal{X},\mathcal{A},\mathcal{P})$ be a majorized experiment. Suppose that $\mathcal{A}$ is the power set of $\mathcal{X}$ and $\mathcal{P}$ contains all the point probability measures. Then the totality of point probability measures D is a maximal orthogonal system of $L_m(\mathcal{E})$. Then the ν_D is the counting measure.

6.2. Maximal orthogonal system and pivotal measure

Let D be a maximal orthogonal system of $L_m(\mathcal{E})$ for a majorized experiment $\mathcal{E}=(\mathcal{X},\mathcal{A},\mathcal{P}=\{P_\theta;\theta\in\Theta\})$. Then ν_D is an equivalent majorizing measure for $\mathcal{E}$. Hence

$$\mathcal{B}_m\equiv \sigma\{dP_\theta/d\nu_D;\theta\in\Theta\}\vee N(\mathcal{P})$$

is a PSS subfield. The next theorem shows that it is the smallest PSS subfield. In particular this shows that this subfield is independent of the choice of a maximal orthogonal system, that is,

$$\sigma\{dP_\theta/d\nu_D;\theta\in\Theta\}\vee N(\mathcal{P}) = \sigma\{dP_\theta/d\nu_G;\theta\in\Theta\}\vee N(\mathcal{P})$$

for any maximal orthogonal systems D and G of $L_m(\mathcal{E})$.

Theorem 6.1. The subfield $\mathcal{B}_m$ is pairwise smallest sufficient and the smallest PSS subfield for $\mathcal{E}$. In particular ν_D is a pivotal measure for each maximal orthogonal system D of $L_m(\mathcal{E})$.

Proof. To prove this, by Corollary 6.1, we can assume $\mathcal{E}=\mathcal{E}^*$ without loss of generality. Obviously $\mathcal{B}_m$ contains a support of P_θ for $\mathcal{E}$ for all $\theta\in\Theta$. Take any θ_1 and θ_2 from Θ, and define μ by $\mu=P_{\theta_1}+P_{\theta_2}$. Then by

$$dP_{\theta_i}/d\mu = dP_{\theta_i}/d\nu_D \cdot I_{[d\mu/d\nu_D>0]} \cdot (d\mu/d\nu_D)^{-1} \text{ a.e. } \mu,$$

P_{θ_i} has a $\mathcal{B}_m$-measurable version of the density w.r.t. μ. This shows that $\mathcal{B}_m$ is pairwise sufficient by Lemma 2.1. Take any pairwise sufficient subfield $\mathcal{C}$. For any $\theta\in\Theta$, by Lemma 6.2(b), there is a countable subset $D(\theta)$ of D such that

$$P_\theta=\sum\{\Pi_w(P_\theta)\,;w\in D(\theta)\}.$$

Let $\mu\in\mathbf{M}(\mathcal{A})$ be a measure of the form $\mu=\sum\{c_w\cdot w;w\in D(\theta)\}$, $c_w>0$. Since $\mathcal{E}=\mathcal{E}^*$ and since in dominated case pairwise sufficiency implies sufficiency, $\mathcal{C}$ is sufficient for

$$\{\|w\|^{-1}w;w\in D(\theta)\}\cup\{\|\Pi_w(P_\theta)\|^{-1}\cdot\Pi_w(P_\theta)\,;\Pi_w(P_\theta)\neq 0,w\in D(\theta)\}.$$

From the proof of Lemma 2.1(2) we may assume that $d(\Pi_w(P_\theta))/d\mu$ and $dw/d\mu$ are $\mathcal{C}$-measurable for all $w\in D(\theta)$. Then

$$dP_\theta/d\nu_D= \sum\{d(\Pi_w(P_\theta))/dw\cdot I_{S(w)}\,;w\in D(\theta)\}$$

$$=\sum\{d(\Pi_w(P_\theta))/d\mu\cdot(dw/d\mu)^{-1}\cdot I_{[dw/d\mu>0]}\,;w\in D(\theta)\} \quad \text{a.e. } \mu,$$

implies that P_θ has a $\mathcal{C}\vee N(\mu)$-measurable version of the density w.r.t. ν_D. Take any θ_1 and $\theta_2\in\Theta$. Then we may suppose that P_{θ_1}, $P_{\theta_2}\ll\mu$ in the above calculation. Therefore $dP_\theta/d\nu_D$ is $\mathcal{C}\vee N(P_{\theta_1}+P_{\theta_2})$-measurable. By the definition of $\tilde{\mathcal{C}}(\mathcal{A})$ we have $\mathcal{B}_m\subset\tilde{\mathcal{C}}(\mathcal{A})$, which implies that $\mathcal{B}_m$ is pairwise smallest sufficient.

If, moreover, $\mathcal{C}$ contains a support of P_θ for all $\theta\in\Theta$, then $[dw/d\nu_D>0]\in\mathcal{C}\vee N(\mathcal{P})$ for all $w\in D$, because, by Proposition 6.1, $W(\mathcal{C})\supset L_m(\mathcal{E})\supset D$ and $\mathcal{C}$ is PSS for the majorized experiment consisting of all probability measures in $W(\mathcal{C})$. And it follows that for all $\theta\in\Theta$

$$dP_\theta/d\nu_D=\sum\{d(\Pi_w(P_\theta))/d\mu\cdot(dw/d\mu)^{-1}I_{[dw/d\nu_D>0]}\,;w\in D(\theta)\} \quad \text{a.e. } \nu_D.$$

From these $dP_\theta/d\nu_D$ is $\mathcal{C}\vee N(\mathcal{P})$-measurable. Hence $\mathcal{B}_m$ is the smallest PSS and ν_D is a pivotal measure for $\mathcal{E}$.

By Theorem 6.1 the measure ν in Example 6.1(a) is a pivotal measure.

To show that each pivotal measure for $\mathcal{E}$ can be represented as the sum of elements of a maximal orthogonal system of $L_m(\mathcal{E})$ we need the following:

Theorem 6.2. Suppose that D is a maximal orthogonal system of $L_m(\mathcal{E})$. Then

$$L_m(\mathcal{E}) = \{f\cdot\nu_D; f\in L^1(\mathcal{X},\mathcal{B}_m,\nu_D/\mathcal{B}_m)\}.$$

Proof. The right hand side includes P by the preceding theorem and is a closed vector sublattice of $\mathfrak{M}(\mathcal{A})$, so it includes $L_m(\mathcal{E})$. For each fixed $w\in D$, applying the method in the proof of Proposition 1.1 in Chapter 1 of Neveu(1975), it follows that for each $f\in L^1(\mathcal{X},\mathcal{B}_m,\nu_D/\mathcal{B}_m)$ there is a $u\in L_m(\mathcal{E})$, such that

$$f\cdot I_{S(w)}=du/d\nu_D\cdot I_{S(w)} \text{ a.e. } \nu_D.$$

Let us denote this u by $u(f,w)$. Since $f\in L^1(\mathcal{X},\mathcal{B}_m,\nu_D/\mathcal{B}_m)$ there exists a countable subset $D_0=\{w_1,w_2,\ldots\}$ of D such that $w([|f|>0])=0$ for all $w\in D-D_0$. Then

$$f\cdot\nu_D=\sum\{f\cdot I_{S(w_i)}\cdot w_i \ ;i\geq1\} = \sum\{u(f,w_i);i\geq1\}\in L_m(\mathcal{E}).$$

Remark 6.1. By Theorem 6.2, $L_m(\mathcal{E})$ is the L-space of the majorized experiment $\mathcal{E}(\mathcal{B}_m)$.

Theorem 6.3. Let n be any pivotal measure for $\mathcal{E}$. Then there exists a maximal orthogonal system G of $L_m(\mathcal{E})$ such that $n=\nu_G$.

Proof. Let D be a maximal orthogonal system of $L_m(\mathcal{E})$ and let $S(w)=[dw/dn>0]$ for all $w\in D$. $S(w)$ is a support of w for $\mathcal{E}$. By Lemma 6.1 dw/dn is $\mathcal{B}_m$-measurable. Hence $S(w)\in\mathcal{B}_m$. There is a countable set $K(w)$ and a pairwise disjoint family $\{S(w,k); k\in K(w)\}$ in $\mathcal{B}_m$ such that $S(w) =\cup\{S(w,k);k\in K(w)\}$ with $0<n(S(w,k))<\infty$ for all $k\in K(w)$. By Theorem 6.2, the measures $I_{S(w,k)}\cdot w$, where $w\in D$ and $k\in K(w)$, also define a maximal orthogonal system of $L_m(\mathcal{E})$. Therefore we assume without any loss of generality that $0<n(S(w))<\infty$ holds for each $w\in D$. Put $v(w)(\cdot)=n(\cdot\cap S(w))$ for all $w\in D$. Then $v(w)\in L_m(\mathcal{E})$ because of $v(w)=((dw/dn)^{-1}\cdot I_{S(w)})\cdot\nu_D$ and Theorem 6.2 since $(dw/dn)^{-1}\cdot I_{S(w)}$ is integrable w.r.t. ν_D. Clearly $v(w)$ and w are equivalent for each $w\in D$. Hence $G=\{v(w);w\in D\}$ is also a maximal orthogonal system of $L_m(\mathcal{E})$. Then it is clear that n and ν_G are equivalent with each other. For any A which is σ-finite w.r.t. n there exists a countable subset $\{w_1,w_2,\ldots.\}$ of D such that $A\subset\bigcup_{i=1}^{\infty}S(w_i)$. Hence we have that $n(A)=\nu_G(A)$. Since both measures n and ν_G have the finite subset property, we have $n=\nu_G$.

Take any $\theta\in\Theta$ and let $S(\theta)$ be a support of P_θ for $\mathcal{E}$ belonging to $\mathcal{B}_m$. Let us consider a maximal decomposition $\mathcal{F}$ of the experiment $\mathcal{E}(\mathcal{B}_m)$. Then $\sum\{P_F(\cdot\cap F);F\in\mathcal{F}\}$ is a pivotal measure for $\mathcal{E}$(Theorem 2.2). This also follows from Theorem 6.1 and the next Theorem 6.4. In Theorem 6.4 the relation between a maximal decomposition of $\mathcal{E}^*(\mathcal{B}_m)$ and a maximal orthogonal system of $L_m(\mathcal{E})$ is shown. Note that each maximal decomposition of $\mathcal{E}(\mathcal{B}_m)$ is a maximal decomposition of $\mathcal{E}^*(\mathcal{B}_m)$.

Theorem 6.4. (a) If $\mathcal{F}$ is a maximal decomposition of $\mathcal{E}^*(\mathcal{B}_m)$ then $\{P_F(\cdot\cap F);F\in\mathcal{F}\}$ is a maximal orthogonal system in $L_m(\mathcal{E})$.
(b) If D is a maximal orthogonal system in $L_m(\mathcal{E})$ and there is a support $S(w)$ of w for $\mathcal{E}$ belonging to $\mathcal{B}_m$ for each $w\in D$, then $\{S(w);w\in D\}$ is a maximal decomposition of $\mathcal{E}^*(\mathcal{B}_m)$.

Proof. (a) Let D be a maximal orthogonal system of $L_m(\mathcal{E})$. By Theorem 6.2, for each $F\in\mathcal{F}$, there exists a $\mathcal{B}_m$-measurable g_F such that

$$P_F(\cdot\cap F) = (I_F\cdot g_F)\cdot\nu_D\in L_m(\mathcal{E}).$$

Clearly $\{P_F(\cdot\cap F);F\in\mathcal{F}\}$ is an orthogonal system in $L_m(\mathcal{E})$. Take any $v\in L_m(\mathcal{E})_+$ such that $v\wedge P_F(\cdot\cap F)=0$ for all $F\in\mathcal{F}$. Again by Theorem 6.9, there exists an $f\in L^1(\mathcal{X},\mathcal{B}_m,\nu_D/\mathcal{B}_m)$ such that $v=f\cdot\nu_D$. Then we have, by $v\wedge P_F(\cdot\cap F)=0$ and $v\sim\nu_D$ on $[f(x)>0]$,

$$0=\nu_D([f(x)>0]\cap[I_F g_F>0])=\nu_D([f(x)>0]\cap F)$$

for all $F\in\mathcal{F}$. Since $\mathcal{F}$ is a maximal decomposition of $(\mathcal{X},\mathcal{B}_m,n)$, where $n(\cdot)=\sum\{P_F(\cdot\cap F);F\in\mathcal{F}\}$, by Lemma 1.4, it follows that $n([f(x)>0])=0$. Since n is equivalent to ν_D we have $\nu_D([f(x)>0])=0$. Hence $v=0$ and this shows $\{P_F(\cdot\cap F);F\in\mathcal{F}\}$ is a maximal orthogonal system of $L_m(\mathcal{E})$.

(b) It is easy to show that $\{S(w);w\in D\}$ is a decomposition of $\mathcal{E}^*(\mathcal{B}_m)$. That is, (1) for any $w_1,w_2\in D$ such that $w_1\neq w_2$, it holds $S(w_1)\cap S(w_2)\in N(\mathcal{P}^*)$, (2) for any $w\in D$ there exists a $w'\in\mathcal{P}^*$ such that $w'(S(w))>0$ and $S(w)-S(w')\in N(\mathcal{P}^*)$. We will show its maximality. Suppose that $B\in\mathcal{B}_m$, $B-S(Q)\in N(\mathcal{P}^*)$ for some $Q\in\mathcal{P}^*$, and that $B\cap S(w)\in N(\mathcal{P}^*)$ for all $w\in D$. By Theorem 6.2, $Q(\cdot\cap B)=(I_B f)\cdot\nu_D$ for some $f\in L^1(\mathcal{X},\mathcal{B}_m,\nu_D/\mathcal{B}_m)$. So we have $Q(\cdot\cap B)\in L_m(\mathcal{E})$ and $Q(\cdot\cap B)\wedge w=0$ for all $w\in D$. Maximality of D implies $Q(\cdot\cap B)=0$. From $Q(B)=0$ and $B\subset S(Q)$ $[\mathcal{P}^*]$, by the definition of support for $\mathcal{E}^*$, we have $B\in N(\mathcal{P}^*)$. This shows that $\{S(w);w\in D\}$ is a maximal decomposition of $\mathcal{E}^*(\mathcal{B}_m)$.

Remark 6.2. Gooßen(1988) and Fujii(1990) obtained similar characterizations of pivotal measure. The following, due to Fujii(1990), is one of them. Let $\mathcal{E}=(\mathcal{X},\mathcal{A},\mathcal{P})$ be weakly dominated and let n be a pivotal measure for $\mathcal{E}$. Then a measure m on $\mathcal{A}$ is pivotal for $\mathcal{E}$ if and only if $m(A)=\int_A f(x)\,dn$ for some $\mathcal{B}$-measurable function such that $P_\theta([f(x)>0])=1$ for all $\theta\in\Theta$, where $\mathcal{B}$ is the smallest sufficient subfield. As an application let $\mathcal{X}=R^k$, $\mathcal{A}$= the Borel field, and let $\mathcal{P}=\{P_\theta;\theta\in\Theta\}$. Here P_θ is the k-th product of $N(\theta,1)$. Then each P_θ is a pivotal measure for $\mathcal{E}$. Any pivotal measure has the density w.r.t. the Lebesgue measure l of the form of $exp\{(-1/2)\cdot\sum_{i=1}^{k} x_i^2\cdot f(\sum_{i=1}^{k} x_i)\}$ for some $f(x)>0$ a.e.l. Hence the Lebesgue measure is pivotal if $k=1$, but is not if $k>1$.

6.3. Bibliographical notes

This chapter is completely indebted to Luschgy, Mussmann and Yamada(1988). As was noted in the text the characterization of the structure of pivotal measures covers the methods of the construction by Ramamoorthi and Yamada (1982), and by Halmos and Savage(1949). Essentially

Proposition 6.2 was already treated by Torgersen(1988).

The notion of minimal L-space was introduced by Le Cam (1964,1986). Torgersen (1977) defined, in dominated case, the minimal L-space by $\{\mu \in L(\mathscr{E})$; μ has a version of the density w.r.t. n in Theorem 5.1 which is measurable w.r.t. the smallest sufficient subfield$\}$. This is the same type of representation as Theorem 6.2. Relationship between the minimal L-space and the smallest sufficient sublattice is found in Torgersen(1991,Proposition 7.3.14). The book of Torgersen also contains a characterization of pairwise sufficiency related to the minimal L-space(Proposition 7.3.7); let $\mathscr{E}=(\mathscr{X},\mathscr{A},\mathscr{P}=\{P_\theta;\theta\in\Theta\})$ be an experiment. Let $\{w_t;t\in T\}$ be a maximal orthogonal system of probability measures in $L_m(\mathscr{E})$. For each $t\in T$ $\mathscr{E}_t$ denotes the dominated experiment $(\mathscr{X},\mathscr{A},\{P_{\theta,t};\theta\in\Theta_t\})$, where $\Theta_t=\{\theta$; the w_t-continuous part $(P_\theta)_t$ of P_θ is not zero$\}$, $P_{\theta,t}=(P_\theta)_t/\|(P_\theta)_t\|$. Then $\mathscr{B}$ is pairwise sufficient if and only if it is pairwise sufficient for $\{w_t;t\in T\}$ as well as for each $\mathscr{E}_t, t\in T$. This would be interesting when it is compared with the result by Fujii and Morimoto(1986) in Section 3.2.

References

Bahadur, R.R. (1954). Sufficiency and statistical decision functions. Ann. Math. Statist. 25, 423-462.

Bahadur, R.R. (1955). A characterization of sufficiency. Ann. Math. Statist. 26, 286-293.

Bahadur, R.R. (1957). On unbiased estimates of uniformly minimum variance. Sankhyā 18, 211-224.

Bahadur, R.R. (1967). A note on UMV estimates and ancillary statistics. In memorial volume dedicated to Hájek. Charles University.

Basu, D. and Ghosh, J.K. (1967). Sufficient statistics in sampling from a finite universe. Proc. 36th Session Int. Stat. Inst., 850-859.

Blackwell, D. (1947). Conditional expectation and unbiased sequential estimation. Ann. Math. Statist. 18, 105-110.

Blackwell, D. (1951). Comparison of experiments. Proc. 2nd Berkeley Symp. on Math. Stat. and Prob., 93-102.

Blackwell, D. (1953). Equivalent conditions of experiments. Ann. Math. Statist. 24, 265-272.

Bomze, I.M. (1990). A functional analytic approach to statistical experiments. Pitman.

Brown, L. (1975). On a theorem of Morimoto concerning sufficiency for discrete distributions. Ann. Statist. 3, 1180-1182.

Burkholder, D.L. (1961). Sufficiency in the undominated case. Ann. Math. Statist. 32, 1191-1200.

Cassel, C., Särndal, C., and Wretman, J.H. (1977). Foundations of Inference in Survey Sampling. John Wiley.

Dawid, A.P. (1980). Conditional independence for statistical operators. Ann. Statist. 8, 598-617.

Diepenbrock, F.R. (1971). Charakteriesierung einer allgemeineren Bedingung als Dominiertheit mit Hilfe von lokalisierten Massen. Thesis. University of Münster.

Fisher, R.A. (1922). On the mathematical foundations of theoretical statistics . Philos. Trans. Roy. Soc.London Ser. A. 222, 309-368.

Fremlin, D.H. (1974). Topological Riesz Space and Measure Theory. Cambridge University Press.

Fremlin, D.H. (1978). Decomposable measure spaces. Z. Wahrscheinlichkeitstheorie. 45, 159-167.

Fujii, J. (1989). On the smallest pairwise sufficient subfield in majorized statistical experiment. Osaka J. Math. 26, 429-446.

Fujii, J. (1990). A note on pivotal measures in majorized experiments. Osaka J. Math. 27, 739-744.

Fujii, J. (1991). Partitions of sample space and smallest pairwise sufficient subfields. Statistics & Decisions. 9, 393-400.

Fujii, J. and Morimoto, H. (1986). Sufficiency and pairwise sufficiency in majorized experiments. Sankhyā Ser.A, 48, 315-330.

Fujii, J. and Morimoto, H. (1993). On Burkholder pathology (in Japanese). Report of Kyodo-Kenkyu of Inst. Stat. Math. 42, 7-10.

Ghosh, J.K., Morimoto, H. and Yamada, S. (1981). Neyman factorization and minimality of pairwise sufficient subfields. Ann. Statist. 9, 514-530.

Ghosh, J.H.(ed.) (1988). Statistical Information and Likelihood. A collection of critical essays by Dr. D. Basu. Lecture notes in Statist. 45. Springer-Verlag.

Gooßen, K. (1988). Partitions and pivotal measures in experiments. Statistics & Decisions. 6, 283-291.

Halmos, P.R. (1950). Measure Theory. van Nostrand.

Halmos, P.R. and Savage, L.J. (1949). Applications of the Radon-Nikodym theorem to the theory of sufficient statistics. Ann. Math. Statist. 20, 225-241.

Hasegawa, M. and Perlman, M. D. (1974). On the existence of a minimal sufficient subfield. Ann. Statist. 2, 1049-1055.
Correction (1975). Ann. Statist. 3, 1371-1372.

Heyer, H. (1982). Theory of statistical experiments. Springer-Verlag.

Heyer,H. and Kusama,T. (1991). Some remarks on weak sufficiency. Statistical Papers 32, 137-154.

Heyer, H and Yamada, S. (1993). Common conditional probability for pairwise sufficiency. (in preparation)

Kakutani, S. (1941). Concrete representation of abstract (L)-space and the mean ergodic theorem. Ann. Math. 42, 523-537.

Kudo, H. (1957). On sufficiency and completeness of statistics (in Japanese). Sugaku 8, 129-138.

Kudo, H. (1972). On structure of statistical spaces (in Japanese). (in Tokeiteki kozō). Sūri Kagaku Kokyuroku

150, 1-22.

Kuratowski,K. (1966). Topology (volume 1). Academic Press.

Kusama, T. (1991). On some class properties of statistical experiments under weak Blackwell equivalence. Tokyo J. Math. 14, 419-424.

Kusama, T. and Yamada, S. (1972). On compactness of the statistical structure and sufficiency. Osaka J. Math. 9, 11-18.

Kusama, T. and Fujii J. (1987). A note on test sufficiency in weakly dominated experiments. Tokyo J. Math. 10. 133-136.

Le Cam, L. (1964). Sufficiency and approximate sufficiency. Ann.Math. Statist. 35, 1419-1455.

Le Cam, L. (1986). Asymptotic methods in statistical decision theory. Springer-Verlag.

Lehmann, E.L. (1983). Theory of Point Estimation. John Wiley.

Lehmann, E.L. and Scheffé, H. (1950). Completeness, similar regions and unbiased estimation. Sankhyā. 10, 305-340.

Littaye, H.,Petit, J., Piednoir L. and van Cutsem, B. (1969). Exhaustivitè, Ann. Inst. H. Poincarè. V, 289-322.

Luschgy, H. (1976). Invariante additive Mengenfunktionen und invariante statistische Testprobleme. Thesis. University of Münster.

Luschgy, H. (1978). Sur l'existence d'une plus petite sous-tribu exhaustive par paire. Ann. Inst. H. Poincarè. Section B. 5, 14, 391-398.

Luschgy, H. (1988). Pairwise sufficiency and invariance. Osaka J. Math. 25, 785-794.

Luschgy,H. (1988). A note on majorized statistical experiments. Sankhyā, Ser. A, 50, 149-150.

Luschgy,H. and Mussmann, D. (1985). Equivalent properties and completion of statistical experiments. Sankhyā Ser. A, 47, 174-195.

Luschgy, H. and Mussmann, D. (1987). A characterization of weakly dominated statistical experiments by compactness of the set of decision rules. Sankhyā, Ser. A. 49, 388-394.

Luschgy, H., Mussmann, D. and Yamada, S. (1988). Minimal L-space and Halmos-Savage criterion for majorized experiments. Osaka J. Math. 25, 795-803.

Martin, F., Petit, J.L. and Littaye, M. (1973). Indépendance conditionnelle dans le modele statistique bayésien. Ann

Inst. H. Poincaré Sect. B, 9, 19-40.

McShane, E.J. (1962). Families of measures and representations of algebras of operators. Trans. Amer. Math. Soc. 102, 328-345.

Michaletzky, G. (1985). A condition to avoid a pathological structure of sufficient σ-fields. Prob. and Math. Statist.5, 153-163.

Michaletzky, G. (1987). On the σ-fields which are larger than a sufficient one. Mathematical Statistics and Probability Theory. A, 231-236.

Morimoto, H. (1972). Statistical structure of the problem of sampling from finite populations. Ann. Math. Statist. 43, 490-497.

Morimoto, H. (1973). Sufficiency in some undominated statistical structures. Indian Statist.Inst.

Mussmann, D. (1971). Schwach dominierte Familien von Massen. Thesis. University of München.

Mussmann, D. (1972). Vergleich von Experimenten in schwach dominierten Fall. Z. Wahrscheinlichkeitstheorie verw. Geb. 24, 295-308.

Mussmann, D. (1983). Generating systems for L-spaces with applications to sufficiency. Proc. 4th Pannonian Symp. on Math. Stat. 277-286.

Mussmann,D.(1991). Conical measures, weak convergence of experiemnts, and minimal sufficiency. Sankhyā, Ser. A, 53, 162-182.

Neveu, J. (1975). Discrete Parameter Martingales. North Holland.

Neyman, J. (1935). Sur un teorema concernente le considette statistiche sufficienti. Giorn. Ist. Ital.Att., 6, 320-334.

Pfanzagl, J. (1974). A characterization of sufficiency by power functions. Metrika. 21, 197-199.

Pfanzagl, J. and Weber, J. (1973). Nonlinear operator and sufficiency. Ann. Statist. 1, 183-187.

Pitcher, T.S. (1957). Sets of measures not admitting necessary and sufficient statistics or subfields. Ann. Math. Statist. 28, 267-268.

Pitcher, T.S. (1965). A more general property than domination for sets of probability measures. Pacific J. Math. 15, 597-611.

Ramamoorthi, R.V. (1980). Sufficiency, pairwise sufficiency

and Bayes sufficiency in undominated experiments. Thesis. Indian Stat.Inst.

Ramamoorthi, R.V. (1980). On sufficiency and pairwise sufficiency in standard Borel spaces. Sankhyā Ser.A, 42, 139-145.

Ramamoorthi, R.V. (1982). Sufficiency and pairwise sufficiency in standard Borel spaces. Osaka J. Math. 19, 577-586.

Ramamoorthi,R.V. and Yamada, S. (1982). Neyman factorization theorems for experiments admitting densities. Sankhyā Ser. A, 45, 168-180.

Ramamoorthi, R.V. and Yamada, S. (1983). On the union of compact statistical structures. Osaka J. Math. 20, 257-264.

Ramamoorthi, R.V. and Roy, K.K. (1987). A note on weakly dominated experiments. Sankhyā, Ser. A, 49, 128-129.

Rao, B.V. and Yamada, S. (1980). Note on a problem of Diepenbrock. Osaka J. Math. 17, 325-327.

Rao, C.R. (1945). Information and accuracy attainable in the estimation of statistical parameters. Bull. Cal. Math. Soc. 37, 81-91.

Rao, K.P.S.B. and Rao B.V. (1981). Borel spaces. Dissertationes Mathematicae.

Rao, M.M. (1975). Conditional measures and operators. J. Multivariate Anal. 5, 330-413.

Rogge, L. (1972). Compactness and domination. Manuscripta Math.7, 299-306.

Roy, K.K. and Ramamoorthi, R.V. (1979). Relationship between Bayes, classical and decision theoretical sufficiency. Sankhyā, Ser. A, 41, 48-58.

Sacksteder, R. (1967). A note on statistical equivalence. Ann. Math. Statist. 38, 787-794.

Schaefer, H.H. (1974). Banach lattices and positive operators. Springer-Verlag.

Siebert, E. (1979a) Pairwise sufficiency. Z. Wahrscheinlichkeitstheorie verw. Geb. 46, 237-246.

Siebert, E. (1979b). Statistical experiments and their conical measures. Z. Wahrscheinlichkeitstheirie verw.Geb. 46, 247-258.

Speed, T.P. (1976). A note on pairwise sufficiency and completions. Sankhyā, Ser. A, 38, 194-196.

Strasser, E. (1985). Mathematical theory of statistics. Walter de Gruyter.

Takeuchi, K. (1969). Unpublished lecture note.

Taylor, A.E. (1958). Introduction to Functional Analysis. John Wiley.

Torgersen, E.N. (1976). Comparison of statistical experiments. Scand. J. Statist. 3, 186-208.

Torgersen, E.N. (1977). Mixtures and products of dominated experiments. Ann. Statist. 5, 44-64.

Torgersen, E.N. (1979). On complete sufficient statistics and uniformly minimum variance unbiased estimation. Sym. Math. Ist. Nazionale Di Alta matematica. 137-152.

Torgersen, E.N. (1988). On Bahadur's converse of the Rao-Blackwell theorem. Extension to majorized experiment. Scand. J. Statist. 15, 273-280.

Torgersen, E.N. (1991). Comparison of Statistical Experiments. Cambridge University Press.

Trenkler, G. (1982). Partitions, sufficiency and undominated families of probability measures. Ann. Inst. Statist. Math.Part A, 34, 151-160.

Yamada, S. (1976). Pivotal measures in the case of weak domination. Osaka J. Math. 18, 621-632.

Yamada, S. (1980). On completions of sigma-fields and pairwise sufficiency. Sankhyā, Ser.A. 42, 185-200.

Yamada, S. (1981). Neyman factorization theorems in the case of non-equivalent dominating measures. Osaka J. Math. 18, 621- 632.

Yamada,S. (1982). On Neyman factorization theorems. (in Japanese). Sugaku 34, 140-151.

Yamada, S.(1985). A note on a representation of a transition by pivotal measure. Osaka J. Math. 22, 299-305.

Yamada, S. (1988). M-space of majorized experiment and pivotal measure. Statistics & Decisions. 6, 163-174.

Yamada, S. (1991). Applications of an extended notion of common conditional probability. Report of Tokyo Univ. Fish. 26, 79-87.

Yamada, S. (1993). Notes on the Burkholder problem. (in preparation).

Yamada, S. and Suzuki, T. (1990). Remarks on Bayes sufficiency. Tokyo J. Math. 13, 147-152.

Yamada, S. and Morimoto, H. (1992). Sufficiency. Current Issues in Statistical Inference: Essays in Honor of D. Basu. IMS Lecture Notes Monograph Series. 17, 86-98.

Zaanen, A.C. (1967). Integration. North Holland.

Subject index

List of symbols

Chapter 1.

6 $N(m)$, $N(Q)$, $P \ll m$, dP/dm 7 $S(\theta)$, $[P(x)]$, Δ, $P \sim m$, $\mathcal{F}$, $\bar{\mathcal{F}}$
8 $\mathcal{A}_\sigma(m)$, $\mathcal{A}(m)$, $m \ll n$, $\mathfrak{M}_1 \ll \mathfrak{M}_2$
9 m/B, $B \cap \mathcal{A}$ 12 a.e. m 13 $A \subset B$ $[m]$, $ess.sup_{P_\theta}$

Chapter 2

21 $\mathcal{B} \subset \mathcal{C}[P]$, $E(I_A/\mathcal{B})$ 22 $\bar{\mathcal{C}}(\mathcal{A})$, $E(I_A/\mathcal{B},m)$ 23 $m/\mathcal{B}$
27 $\bar{\mathcal{A}}(P_\theta)$, $\tilde{\mathcal{A}}(m)$ 32 $\hat{\mathcal{B}}(\mathcal{A})$

Chapter 3

39 $\mathcal{B}(\mathcal{T})$, $\mathcal{T} > \mathcal{U}$ 41 $\mathcal{T}(\mathcal{B})$, $x \sim y$ 43 P_x, $\mathfrak{M}$ 44 π_x,
$\sigma(\mathcal{T})$ 47 $\sigma(M)$ 48 $\tilde{\mathcal{C}}(\mathcal{A})$ 49 $\mathcal{S}_0$, $\mathcal{S}$, $\Theta(F)$, $P(F)$, $\mathcal{E}(F)$
51 $A \in \mathcal{B}[P]$, $A = B$ $[P]$ 53 $\hat{\mathcal{B}}(\mathcal{B})$, $\tilde{\mathcal{B}}(\mathcal{B},)$ $A \subset B$ $[P]$ 57 $\bar{\bar{\mathcal{B}}}(\mathcal{B})$

Chapter 4

59 $L^2(\mathcal{X},\mathcal{A},P_\theta)$ 63 $L^1(\mathcal{X},\mathcal{A},P_\theta)$ 71 U, T, $\tilde{T}$, N, $\mathcal{B}$

Chapter 5

78 $\mathfrak{M}(\mathcal{A})$, $L(\mathcal{E})$, $M(\mathcal{E})$, $\delta(\mathcal{E}(\mathcal{B}),\mathcal{E})$ 78 $f.n$ 79 $L(K)$, $L(n)$
81 T_n 82 $m(\mathcal{A})$ $b(\mathcal{A}(m))$ $\|b(\mathcal{A}(m))\|$ 83
$(M)-\int b(\mathcal{A}(m))\, dm$, 84 $(T)-\int f(P)\, d\mu$ 85 $U(\mathcal{E})$, $V(\mathcal{E})$
93 $\mathcal{F}(\mathcal{A})$, $\rho(f)$, T_n' 102 $T_n(\mathcal{C})$, $T_n'(\mathcal{C})$ 107 $f_A(P)$

Chapter 6

108 $L_m(\mathcal{E})$, $\mathcal{E}^*$, $W(\mathcal{B})$ 110 $\sigma(H)$, $\mathcal{S}_1 \vee \mathcal{S}_2$ 111 $\Pi_x(y)$
112 ν_D, $B(D)$ 113 $\mathcal{B}_m$ 116 $v(w)$ 117 $N(\theta,1)$

Vol. 1: R.A. Fisher: An Appreciation. Edited by S.E. Fienberg an D.V. Hinkley. XI, 208 pages, 1980.

Vol. 2: Mathematical Statistics and Probability Theory. Proceedings 1978. Edited by W. Klonecki, A. Kozek, and J. Rosinski. XXIV, 373 pages, 1980.

Vol. 3: B.D. Spencer, Benefit-Cost Analysis of Data Used to Allocate Funds. VIII, 296 pages, 1980.

Vol. 4: E.A. van Doorn, Stochastic Monotonicity and Queueing Applications of Birth-Death Processes. VI, 118 pages, 1981.

Vol. 5: T. Rolski, Stationary Random Processes Associated with Point Processes. VI, 139 pages, 1981.

Vol. 6: S.S. Gupta and D.-Y. Huang, Multiple Statistical Decision Theory: Recent Developments. VIII, 104 pages, 1981.

Vol. 7: M. Akahira and K. Takeuchi, Asymptotic Efficiency of Statistical Estimators. VIII, 242 pages, 1981.

Vol. 8: The First Pannonian Symposium on Mathematical Statistics. Edited by P. Révész, L. Schmetterer, and V.M. Zolotarev. VI, 308 pages, 1981.

Vol. 9: B. Jørgensen, Statistical Properties of the Generalized Inverse Gaussian Distribution. VI, 188 pages, 1981.

Vol. 10: A.A. McIntosh, Fitting Linear Models: An Application of Conjugate Gradient Algorithms. VI, 200 pages, 1982.

Vol. 11: D.F. Nicholls and B.G. Quinn, Random Coefficient Autoregressive Models: An Introduction. V, 154 pages, 1982.

Vol. 12: M. Jacobsen, Statistical Analysis of Counting Processes. VII, 226 pages, 1982.

Vol. 13: J. Pfanzagl (with the assistance of W. Wefelmeyer), Contributions to a General Asymptotic Statistical Theory. VII, 315 pages, 1982.

Vol. 14: GLIM 82: Proceedings of the International Conference on Generalised Linear Models. Edited by R. Gilchrist. V, 188 pages, 1982.

Vol. 15: K.R.W. Brewer and M. Hanif, Sampling with Unequal Probabilities. IX, 164 pages, 1983.

Vol. 16: Specifying Statistical Models: From Parametric to Non-Parametric, Using Bayesian or Non-Bayesian Approaches. Edited by J.P. Florens, M. Mouchart, J.P. Raoult, L. Simar, and A.F.M. Smith, XI, 204 pages, 1983.

Vol. 17: I.V. Basawa and D.J. Scott, Asymptotic Optimal Inference for Non-Ergodic Models. IX, 170 pages, 1983.

Vol. 18: W. Britton, Conjugate Duality and the Exponential Fourier Spectrum. V, 226 pages, 1983.

Vol. 19: L. Fernholz, von Mises Calculus For Statistical Functionals. VIII, 124 pages, 1983.

Vol. 20: Mathematical Learning Models — Theory and Algorithms: Proceedings of a Conference. Edited by U. Herkenrath, D. Kalin, W. Vogel. XIV, 226 pages, 1983.

Vol. 21: H. Tong, Threshold Models in Non-linear Time Series Analysis. X, 323 pages, 1983.

Vol. 22: S. Johansen, Functional Relations, Random Coefficients and Nonlinear Regression with Application to Kinetic Data, VIII, 126 pages, 1984.

Vol. 23: D.G. Saphire, Estimation of Victimization Prevalence Using Data from the National Crime Survey. V, 165 pages, 1984.

Vol. 24: T.S. Rao, M.M. Gabr, An Introduction to Bispectral Analysis and Bilinear Time Series Models. VIII, 280 pages, 1984.

Vol. 25: Time Series Analysis of Irregularly Observed Data. Proceedings, 1983. Edited by E. Parzen. VII, 363 pages, 1984.

Vol. 26: Robust and Nonlinear Time Series Analysis. Proceedings, 1983. Edited by J. Franke, W. Härdle and D. Martin. IX, 286 pages, 1984.

Vol. 27: A. Janssen, H. Milbrodt, H. Strasser, Infinitely Divisible Statistical Experiments. VI, 163 pages, 1985.

Vol. 28: S. Amari, Differential-Geometrical Methods in Statistics. V, 290 pages, 1985.

Vol. 29: Statistics in Ornithology. Edited by B.J.T. Morgan and P.M. North. XXV, 418 pages, 1985.

Vol 30: J. Grandell, Stochastic Models of Air Pollutant Concentration. V, 110 pages, 1985.

Vol. 31: J. Pfanzagl, Asymptotic Expansions for General Statistical Models. VII, 505 pages, 1985.

Vol. 32: Generalized Linear Models. Proceedings, 1985. Edited by R. Gilchrist, B. Francis and J. Whittaker. VI, 178 pages, 1985.

Vol. 33: M. Csörgo, S. Csörgo, L. Horváth, An Asymptotic Theory for Empirical Reliability and Concentration Processes. V, 171 pages, 1986.

Vol. 34: D.E. Critchlow, Metric Methods for Analyzing Partially Ranked Data. X, 216 pages, 1985.

Vol. 35: Linear Statistical Inference. Proceedings, 1984. Edited by T. Calinski and W. Klonecki. VI, 318 pages, 1985.

Vol. 36: B. Matérn, Spatial Variation. Second Edition. 151 pages, 1986.

Vol. 37: Advances in Order Restricted Statistical Inference. Proceedings, 1985. Edited by R. Dykstra, T. Robertson and F.T. Wright. VIII, 295 pages, 1986.

Vol. 38: Survey Research Designs: Towards a Better Understanding of Their Costs and Benefits. Edited by R.W. Pearson and R.F. Boruch. V, 129 pages, 1986.

Vol. 39: J.D. Malley, Optimal Unbiased Estimation of Variance Components. IX, 146 pages, 1986.

Vol. 40: H.R. Lerche, Boundary Crossing of Brownian Motion. V, 142 pages, 1986.

Vol. 41: F. Baccelli, P. Brémaud, Palm Probabilities and Stationary Queues. VII, 106 pages, 1987.

Vol. 42: S. Kullback, J.C. Keegel, J.H. Kullback, Topics in Statistical Information Theory. IX, 158 pages, 1987.

Vol. 43: B.C. Arnold, Majorization and the Lorenz Order: A Brief Introduction. VI, 122 pages, 1987.

Vol. 44: D.L. McLeish, Christopher G. Small, The Theory and Applications of Statistical Inference Functions. VI, 124 pages, 1987.

Vol. 45: J.K. Ghosh (Ed.), Statistical Information and Likelihood. 384 pages, 1988.

Vol. 46: H.-G. Müller, Nonparametric Regression Analysis of Longitudinal Data. VI, 199 pages, 1988.

Vol. 47: A.J. Getson, F.C. Hsuan, {2}-Inverses and Their Statistical Application. VIII, 110 pages, 1988.

Vol. 48: G.L. Bretthorst, Bayesian Spectrum Analysis and Parameter Estimation. XII, 209 pages, 1988.

Vol. 49: S.L. Lauritzen, Extremal Families and Systems of Sufficient Statistics. XV, 268 pages, 1988.

Vol. 50: O.E. Barndorff-Nielsen, Parametric Statistical Models and Likelihood. VII, 276 pages, 1988.

Vol. 51: J. Hüsler, R.-D. Reiss (Eds.), Extreme Value Theory. Proceedings, 1987. X, 279 pages, 1989.

Vol. 52: P.K. Goel, T. Ramalingam, The Matching Methodology: Some Statistical Properties. VIII, 152 pages, 1989.

Vol. 53: B.C. Arnold, N. Balakrishnan, Relations, Bounds and Approximations for Order Statistics. IX, 173 pages, 1989.

Vol. 54: K.R. Shah, B.K. Sinha, Theory of Optimal Designs. VIII, 171 pages, 1989.

Vol. 55: L. McDonald, B. Manly, J. Lockwood, J. Logan (Eds.), Estimation and Analysis of Insect Populations. Proceedings, 1988. XIV, 492 pages, 1989.

Vol. 56: J.K. Lindsey, The Analysis of Categorical Data Using GLIM. V, 168 pages, 1989.

Vol. 57: A. Decarli, B.J. Francis, R. Gilchrist, G.U.H. Seeber (Eds.), Statistical Modelling. Proceedings, 1989. IX, 343 pages, 1989.

Vol. 58: O.E. Barndorff-Nielsen, P. Blæsild, P.S. Eriksen, Decomposition and Invariance of Measures, and Statistical Transformation Models. V, 147 pages, 1989.

Vol. 59: S. Gupta, R. Mukerjee, A Calculus for Factorial Arrangements. VI, 126 pages, 1989.

Vol. 60: L. Györfi, W. Härdle, P. Sarda, Ph. Vieu, Nonparametric Curve Estimation from Time Series. VIII, 153 pages, 1989.

Vol. 61: J. Breckling, The Analysis of Directional Time Series: Applications to Wind Speed and Direction. VIII, 238 pages, 1989.

Vol. 62: J.C. Akkerboom, Testing Problems with Linear or Angular Inequality Constraints. XII, 291 pages, 1990.

Vol. 63: J. Pfanzagl, Estimation in Semiparametric Models: Some Recent Developments. III, 112 pages, 1990.

Vol. 64: S. Gabler, Minimax Solutions in Sampling from Finite Populations. V, 132 pages, 1990.

Vol. 65: A. Janssen, D.M. Mason, Non-Standard Rank Tests. VI, 252 pages, 1990.

Vol. 66: T. Wright, Exact Confidence Bounds when Sampling from Small Finite Universes. XVI, 431 pages, 1991.

Vol. 67: M.A. Tanner, Tools for Statistical Inference: Observed Data and Data Augmentation Methods. VI, 110 pages, 1991.

Vol. 68: M. Taniguchi, Higher Order Asymptotic Theory for Time Series Analysis. VIII, 160 pages, 1991.

Vol. 69: N.J.D. Nagelkerke, Maximum Likelihood Estimation of Functional Relationships. V, 110 pages, 1992.

Vol. 70: K. Iida, Studies on the Optimal Search Plan. VIII, 130 pages, 1992.

Vol. 71: E.M.R.A. Engel, A Road to Randomness in Physical Systems. IX, 155 pages, 1992.

Vol. 72: J.K. Lindsey, The Analysis of Stochastic Processes using GLIM. VI, 294 pages, 1992.

Vol. 73: B.C. Arnold, E. Castillo, J.-M. Sarabia, Conditionally Specified Distributions. XIII, 151 pages, 1992.

Vol. 74: P. Barone, A. Frigessi, M. Piccioni, Stochastic Models, Statistical Methods, and Algorithms in Image Analysis. VI, 258 pages, 1992.

Vol. 75: P.K. Goel, N.S. Iyengar (Eds.), Bayesian Analysis in Statistics and Econometrics. XI, 410 pages, 1992.

Vol. 76: L. Bondesson, Generalized Gamma Convolutions and Related Classes of Distributions and Densities. VIII, 173 pages, 1992.

Vol. 77: E. Mammen, When Does Bootstrap Work? Asymptotic Results and Simulations. VI, 196 pages, 1992.

Vol. 78: L. Fahrmeir, B. Francis, R. Gilchrist, G. Tutz (Eds.), Advances in GLIM and Statistical Modelling: Proceedings of the GLIM92 Conference and the 7th International Workshop on Statistical Modelling, Munich, 13-17 July 1992. IX, 225 pages, 1992.

Vol. 79: N. Schmitz, Optimal Sequentially Planned Decision Procedures. XII, 209 pages, 1992.

Vol. 80: M. Fligner, J. Verducci (Eds.), Probability Models and Statistical Analyses for Ranking Data. XXII, 306 pages, 1992.

Vol. 81: P. Spirtes, C. Glymour, R. Scheines, Causation, Prediction, and Search. XXIII, 526 pages, 1993.

Vol. 82: A. Korostelev and A. Tsybakov, Minimax Theory of Image Reconstruction. XII, 268 pages, 1993.

Vol. 83: C. Gatsonis, J. Hodges, R. Kass, N. Singpurwalla (Eds.), Case Studies in Bayesian Statistics. XII, 437 pages, 1993.

Vol. 84: S. Yamada, Pivotal Measures in Statistical Experiments and Sufficiency. VII, 129 pages, 1994.